Lecture Notes in Mathematics

A collection of informal reports and seminars
Edited by A. Dold, Heidelberg and B. Eckmann, Zürich

73

Pierre E. Conner
University of Virginia, Charlottesville

Lectures on the Action of a Finite Group

1968

Springer-Verlag Berlin · Heidelberg · New York

Library of Congress Catalog Card Number 68-57939 Printed in Germany. Title No. 3679

CONTENTS

INTRODUCTION

These notes are based upon a series of lectures given, by the kind invitation of Professor Albrecht Dold, at the Mathematics Institute of the University of Heidelberg. The first chapter is aimed at a demonstration of the principle of borrowing ideas and techniques from the various branches of modern algebraic topology and using them to attack a problem in transformation groups. As far as we know, this principle was formally stated first by Borel, who followed it in a most elegant fashion.

We simply contrive a problem and then set about it. In defense of the question we should point out that there has been recently a considerable interest, with profitable results, in the application of vector bundles with operators to the study of finite transformation groups. The reader need only refer to [2, 11]. We shall restrict our attention to complex line bundles with operators. Thus (π, X) is a left action of the finite group π on a space, X. A line bundle with operators $(\pi, \zeta) \longrightarrow (\pi, X)$ is a line bundle over X together with a left action of π on ζ as a group of complex linear bundle maps covering the action of π on X. Via the tensor product these form an abelian group $\mathcal{L}(\pi, X)$. The unit is the action of π on the product bundle given by $h(x, z) = (hx, z)$, $h \in \pi$. We ask how $\mathcal{L}(\pi, X)$ is determined.

If we think of the case of line bundles alone, without operators, we immediately recall that such is determined uniquely by its Chern class in $H^2(X;Z)$. Our first idea then is to find a suitable replacement, a kind of equivariant cohomology group, in which we can associate to every line bundle with operators a Chern class which uniquely determines $(\Pi, \xi) \longrightarrow (\Pi, X)$. A direct obstruction approach does not prove satisfactory, but another route is at hand. Let $\mathcal{C}^* \rightarrow X$ be the sheaf of continuous maps into C^*, the abelian multiplicative group of non-zero complex numbers. Then $H^1(X; \mathcal{C}^*)$ is first shown to be isomorphic to the group of complex line bundles over X, and next it is proved that $H^1(X; \mathcal{C}^*) \simeq H^2(X;Z)$.

The identification of $H^1(X; \mathcal{C}^*)$ is possible because a line bundle arises from a co-ordinate bundle and the defining equation for a co-ordinate system in an abelian structure group is really a cocycle condition and equivalent co-ordinate systems are cohomologous. Our first task then is to define and study co-ordinate systems with operators and this is carried out in section 1. We cover X by $\mathcal{V}$ an indexed collection of open Π-invariant sets, then a co-ordinate system with operators is an assignment to each pair (i,j) of a map

$$r_{j,i} : \Pi \times (V_j \cap V_i) \longrightarrow C^*$$

so that for $x \in V_k \cap V_j \cap V_i$ and h_1, h_2 in Π

$$r_{k,i}(h_1 h_2, x) = r_{k,j}(h_1, h_2 x)\, r_{j,i}(h_2, x).$$

If $e \in \pi$ is the identity then it follows $\{r_{j,i}(e,x)\}$ is a co-ordinate system in the usual sense. On each $V_j \times C$ an action of π is defined by

$$h(x,z) = (hx, r_{j,j}(h,x)z)$$

and the formula

$$r_{j,j}(h,x)r_{j,i}(e,x) = r_{j,i}(h,x) = r_{j,i}(e,hx)r_{i,i}(h,x)$$

on $V_j \cap V_i$ guarantees compatibility of these local actions. A co-ordinate system with operators is "trivial" if we can find map $\lambda_i : V_i \longrightarrow C^*$ so that $r_{j,i}(h,x) = \lambda_i(x)/\lambda_j(hx)$ on $V_j \cap V_i$. Thus we have made a simple extension of the $g_{j,i}$ in Steenrod's book.

In what sense is the resulting group of equivalence classes of co-ordinate systems with operators a 1-dimensional cohomology? The appropriate definitions are discussed in sections 2 and 3. Let K denote the nerve of $\mathcal{V}$. Then a covariant stack with operators on K is defined as follows. To each simplex, $\sigma \subset K$, associate $\mathcal{C}^*(\sigma)$ the multiplicative abelian group of all maps $f : \operatorname{Sup}(\sigma) \longrightarrow C^*$. This is given a right $Z(\pi)$-module structure by $(fh)(x) = f(hx)$. If $\sigma' \subset \sigma$ then $\operatorname{Sup}(\sigma) \subset \operatorname{Sup}(\sigma')$ and restriction is the $Z(\pi)$-module homomorphism $\mathcal{C}^*(\sigma' \to \sigma) : \mathcal{C}^*(\sigma') \to \mathcal{C}^*(\sigma)$. Consider a function φ which assigns to each triple $(V_i, V_j; h)$ a function $\varphi(V_i, V_j; h) \in \mathcal{C}^*(V_j \cap V_i)$. Write this function as $r_{j,i}(h,x)$. Now $\delta\varphi$ should be evaluated on a $(V_k, V_j, V_i; h_1, h_2)$ and omitting the connecting homomorphisms we should put

$$\delta\varphi(V_k,V_j,V_i;h_1,h_2) = \varphi(V_j,V_i;h_2)\varphi(V_k,V_i;h_1h_2)^{-1}\varphi(V_k,V_j;h_1)h_2.$$

The last term is meaningful because $\mathcal{G}(V_k \cap V_j \cap V_i)$ is a right $Z(\pi)$-module. Considering the definition of the action of h_2 the coboundary formula is equivalent to

$$r_{j,i}(h_2,x)r_{k,i}(h_1h_2,x)^{-1}r_{k,j}(h_1,h_2x)$$

thus φ is a 1-cocycle if and only if

$$r_{k,i}(h_1h_2,x) = r_{k,j}(h_1,h_2x)r_{j,i}(h_2,x).$$

Now to give a meaning to this we next turn to homological algebra. The reader will recognize that we follow MacLane's book closely. The setting is a simplicial complex K together with a covariant stack $\mathcal{A} \rightarrow K$ assigning to each simplex σ a right $Z(\pi)$-module and to each $\sigma' \subset \sigma$ a $Z(\pi)$-module homomorphism $\mathcal{A}(\sigma' \rightarrow \sigma) : \mathcal{A}(\sigma') \rightarrow \mathcal{A}(\sigma)$. We want to define $H^p(\pi;\mathcal{A})$, the cohomology of the group π with coefficients in a covariant stack with operators. We put together the usual simplicial definition of cohomology with coefficients in a stack and the definition by means of the (un-normalized) bar construction, of the cohomology of a group, π, with coefficient in a (right) $Z(\pi)$-module. It comes out like this. A <u>basic</u> p-chain, b_p, is a pair $(V^o,\ldots,V^p;h_1,\ldots,h_2)$ where $V^o,\ldots,V^p$ is an ordered (p+1)-tuple of vertices spanning a simplex $|b_p|$ in K and $h_1,\ldots,h_p$ is an ordered p-tuple of elements of π. A basic 0-chain is a pair $(V^o;\ (\ ))$ where $(\ )$ generates $\beta_o(\pi)$ a free right $Z(\pi)$-module. A p-cocycle $\varphi \in C^p(\pi;\mathcal{A})$ is a function which to each basic p-chain, b_p, assigns a value $\varphi(b_p) \in \mathcal{A}(|b_p|)$. Omitting the connecting homomorphisms

$$\delta\varphi(b_{p+1}) = \varphi(V^1,\dots,V^{p+1};h_2,\dots,h_{p+1})$$
$$+\sum_{j=1}^{p} (-1)^j \varphi(V^o,\dots,\hat{V}^j,\dots,V^{p+1};h_1,\dots,h_j h_{j+1},\dots,h_{p+1})$$
$$+ (-1)^{p+1}\varphi(V^o,\dots,V^p;h_1,\dots,h_p)h_{p+1}.$$

This defines $H^p(K;\mathcal{A})$ and in fact $H^1(K;\mathcal{G}^*)$ in the situation discussed previously is just the group of equivalences classes of co-ordinate systems with operators on the covering $\mathcal{V}$. This precise definition is suggested by Bredon's treatment of equivariant cohomology

In this form it appears difficult to compute $H^p(\pi;\mathcal{A})$. However, think of it this way. There is a covariant functor $\sigma \to C(\sigma)$ assigning to each simplex the simplicial chain complex of that simplex. Each $C(\sigma)$ then is a simplicial module equipped with face and degeneracy operators, [10,p.233]. Likewise, the bar construction $\beta(\pi)$ has face and degeneracy operators and is acted on by π as a group of "simplicial maps." Then for each simplex σ we formed the simplicial product

$$C(\sigma) \times \beta(\pi) = \{C_p(\sigma) \otimes_Z \beta_p(\pi)\}.$$

We then used the simplicial structure to define our coboundary. Clearly we could have formed a tensor product of chain complexes

$$C(\sigma) \otimes_Z \beta(\pi) = \{ \sum_{r+s=p} C_r(\sigma) \otimes_Z \beta_s(\pi) \}.$$

This would amount to using $b_{r,s}$-chains, $(v^o,\ldots,v^r;h_1,\ldots,h_s)$ with $|b_{r,s}|$ the simplex spanned by $(v^o,\ldots,v^r)$. Then $\varphi \in C^{r,s}(\pi;\mathcal{A})$ assigns to $b_{r,s}$ a value $\varphi(b_{r,s}) \in \mathcal{A}(|b_{r,s}|)$. Then $C^p(\pi;\mathcal{A}) = \sum_{r+s=p} C^{r,s}(\pi;\mathcal{A})$ becomes a bigraded complex with $\delta = \delta' + \delta''$ defined in the obvious way. It is the Eilenberg-Zilber Theorem which shows that this bigraded cochain complex also yields $H^p(\pi;\mathcal{A})$. The equivalence in Eilenberg-Zilber is natural, so it carries the action of π as a group of simplicial maps.

Now we have two spectral sequences for $H^*(\pi;\mathcal{A})$. The $''E_2^{s,r}$-term is $H^s(\pi;H^r(K;\mathcal{A}))$, the cohomology of π with coefficients in the right $Z(\pi)$-module, $H^r(K;\mathcal{A})$. The $'E_2^{r,s}$ is $H^r(K;\mathcal{R}^s)$ where $\mathcal{R}^s \longrightarrow K$ is the stack $\sigma \longrightarrow H^s(\pi;\mathcal{A}(\sigma))$.

We have arrived, then, by a semi-algebraic route in well-known territory. For what have we but the spectral sequences introduced by Cartan and Borel, [5, p. 151]. If each $\mathcal{A}(\sigma)$ is free, then the 'E-spectral sequence shows $H^p(\pi;\mathcal{A}) \simeq H^p(K;\mathcal{R}^o) = H^p(K;\mathcal{A}^\pi)$ so $''E_t^{s,r} \Rightarrow H^*(K;\mathcal{A}^\pi)$ with $''E_2^{s,r} \simeq H^s(\pi;H^r(K;\mathcal{A}))$. This is the analogue of the spectral sequence of a covering (i.e. π acts freely on X). In fact, Bredon's treatment of this area in [5, p. 151] is the same as ours except that he uses a free acyclic space for π rather than the bar construction. Our approach so far is along the lines of Čech theory and this is suited to our interest.

In section 6 we come down to sheaves with operators and we turn aside from the Čech approach. We take π to act trivially on the base space Y so a sheaf with operators $(\mathcal{A},\pi) \to Y$ is just a sheaf together with a right action of π as a group of stalk preserving sheaf automorphisms. There are two ways to define $H^p(\pi;\mathcal{A})$. We could note that the canonical resolution has natural operators

$$0 \to (\mathcal{A},\pi) \to (\mathcal{C}^o(Y;\mathcal{A}),\pi) \to \ldots \to (\mathcal{C}^r(Y;\mathcal{A}),\pi) \to \ldots.$$

so that the global sections

$$C^r(Y;\mathcal{A}) = \Gamma(Y;\mathcal{C}^r(Y;\mathcal{A}))$$

receive a right $Z(\pi)$-module structure. Let $C^{r,s}(\pi;\mathcal{A}) = \mathrm{Hom}_{Z(\pi)}(\beta_s(\pi), C^r(Y;\mathcal{A}))$, where $\{\beta_s(\pi)\}$ is the bar construction and then take $H^p(\pi;\mathcal{A})$ to be the cohomology of the resulting bigraded cochain complex. We immediately see this way that $"E_2^{s,r} \simeq H^s(\pi;H^r(Y;\mathcal{A}))$. There is a second approach which is to associate with $(\mathcal{A},\pi)$ a certain differential sheaf $\Lambda(\mathcal{A}) : \Lambda^o(\mathcal{A}) \xrightarrow{\delta''} \ldots \xrightarrow{\delta''} \Lambda^s(\mathcal{A}) \to \ldots$ where $\Gamma(U,\Lambda^s(\mathcal{A})) \equiv \mathrm{Hom}_{Z(\pi)}(\beta_s(\pi),\Gamma(U,\mathcal{A}))$ for every open U. When the canonical resolution of a differential sheaf is taken there always results a bigraded cochain complex, [5]. We take $H^p(\pi;\mathcal{A})$ to be the cohomology of this bigraded complex. This time it is clear that $'E_2^{r,s} \simeq H^r(Y;\mathcal{H}^s)$ where $\Gamma(U,\mathcal{H}^s) \equiv H^s(\pi;\Gamma(U,\mathcal{A}))$.

The proof of the equivalence of the two definitions, both of which have desirable features, proceeds as follows. Take any acyclic resolution by sheaves with operators

$$0 \to (\mathcal{A},\pi) \to (\mathcal{L}^0,\pi) \to \dots \to (\mathcal{L}^j,\pi) \to \dots .$$

Apply the functor Λ to get

$$\begin{array}{ccccccc}
\Lambda^0(\mathcal{A}) & \to \Lambda^0(\mathcal{L}^0) & \to \dots & \to \Lambda^0(\mathcal{L}^j) & \to \dots\dots \\
\downarrow & \downarrow & & \downarrow & \\
\Lambda^1(\mathcal{A}) & \to \Lambda^1(\mathcal{L}^0) & \to \dots & \to \Lambda^1(\mathcal{L}^j) & \to \dots \\
\downarrow & \downarrow & & \downarrow & \\
\vdots & \vdots & & \vdots & \\
\Lambda^i(\mathcal{A}) & \to \Lambda^i(\mathcal{L}^0) & \to \dots & \to \Lambda^i(\mathcal{L}^j) & \to
\end{array}$$

Each horizontal row is still an acyclic resolution and vertical arrows are coboundary operators dual to $\partial : \beta_i(\pi) \to \beta_{i-1}(\pi)$. We can compress this into a bigraded differential sheaf

$$\Lambda(\mathcal{L})^s = \sum_{i+j=s} \Lambda^i(\mathcal{L}^j)$$

As $\Lambda(\mathcal{L})$ is a differential sheaf, its canonical resolution yields a bigraded cochain complex $K^{r,s} = C^r(Y;\Lambda(\mathcal{L})^s)$, which is secretly trigraded of course. We first show $C^{r,s}(\pi;\mathcal{A}) = C^r(Y,\Lambda^s(\mathcal{A})) \subset K^{r,s}$ induces an isomorphism of the second definition

$$H^p(\pi,\mathcal{A}) \simeq H^p(K).$$

The "E-spectral sequence for K then shows $H^p(K) \simeq H^p(\sum_{r+s=p} \mathrm{Hom}_{Z(\pi)}(\beta_s(\pi),\Gamma(Y,\mathcal{L}^r))$.

Incidentally, in section 6 we also show that to each exact sequence of sheaves with operators

$$0 \to (\mathcal{A}', \Pi) \to (\mathcal{A}, \Pi) \to (\mathcal{A}'', \Pi) \to 0$$

there is associated an exact coefficient sequence in cohomology,

$$\ldots \to H^p(\Pi; \mathcal{A}') \to H^p(\Pi; \mathcal{A}) \to H^p(\Pi; \mathcal{A}'') \to \ldots \; .$$

This shows us that we shall be able to get satisfactory Chern classes for our line bundles with operators.

In section 7 we make a few remarks about the Cech definition of $H^p(\Pi; \mathcal{A})$ for a sheaf with operators, but with one exception we do not really go into it at all. By this time, the reader will see a Čech definition can be done and how to make it come out to agree with section 6. Still, the Čech description is the whole point in relating sheaves with operators to bundles with operators.

In section 8 we treat sheaves with operators $(\mathcal{A}, \Pi) \to (\Pi, X)$ where Π is not trivial on the base. To each $h \in \Pi$ there is associated a sheaf cohomomorphism

$$\begin{array}{ccc} \mathcal{A} & \xleftarrow{h^\#} & \mathcal{A} \\ \downarrow & & \downarrow \\ X & \xrightarrow{h} & X \end{array}$$

satisfying appropriate and obvious composition rules. Now we could take the direct image sheaf $\mathcal{A}^* \to X/\Pi$ over the quotient space. It is induced by the quotient map $\gamma: X \longrightarrow X/\pi$ and is a sheaf with operators $(\mathcal{A}^*, \pi) \longrightarrow X/\pi$. We could say $H^p(\pi;\mathcal{A}) = H^p(\pi;\mathcal{A}^*)$, turning back, thereby, to section 6, or we could take the canonical resolution with operators, note $C^r(X;\mathcal{A})$ is a $Z(\pi)$-module, and put $C^{r,s}(\pi;\mathcal{A}) = \mathrm{Hom}_{Z(\pi)}(\beta_s(\pi), C^r(X;\mathcal{A}))$ and pass to the cohomology of this bigraded cochain complex. The definitions are equivalent. From the second we immediately see $"E_2^{s,r} \simeq H^s(\pi;H^r(X;\mathcal{A}))$. In section 8 there are some useful technical remarks about the 'E-spectral sequence for $H^*(\pi;\mathcal{A})$.

If $\mathcal{C}^* \longrightarrow X$ is the sheaf of germs of maps into C^* then composition makes $(\mathcal{C}^*, \pi) \longrightarrow (\pi, X)$ a sheaf with operators and $H^1(\pi; \mathcal{C}^*)$ is the group of equivalence classes of complex line bundles with operators over (π, X). Now let π operate on the constant sheaf $X \times Z$ by $(x,n)h = (h^{-1}x,n)$ and denote this sheaf with operators by $(\mathcal{Z}, \pi) \longrightarrow (\pi, X)$. In section 9 we then prove

$$H^1(\pi;\mathcal{C}^*) \simeq H^2(\pi;\mathcal{Z}).$$

So we have the Chern classes for line bundles with operators and for $H^*(\pi;\mathcal{Z})$ we have

$$"E_2^{s,r} \simeq H^s(\pi;H^r(X;Z)).$$

In sections 4 and 10 we discuss holomorphic line bundles over (π, V^n) which is a group of holomorphic isometries in a Kähler metric on the closed connected complex analytic manifold V^n.

The reader will see that our approach is especially well adapted to this situation. Denote by $\mathcal{P}(\pi, V^n)$ the Picard group of those holomorphic line bundles with operators which are topologically equivalent to the trivial line bundle with operators $h(x, \mathfrak{z}) = (hx, \mathfrak{z})$. We can just follow the Kodaira-Spencer treatment of the standard Picard group $\mathcal{P}(V^n)$, [15] to determine $\mathcal{P}(\pi, V^n)$ and show it is a closed connected analytic subgroup of $\mathcal{P}(V^n)$.

The reader will recognize that $H^*(\pi; \mathfrak{z})$ is a simple example of an equivariant cohomology theory as discussed by Bredon in [4, Sec. 5]. As such, it has applications other than to the classification of line bundles with operators. For example, if $\pi = Z_p$, a cyclic group of prime order then it was the study of $H^*(Z_p; \mathfrak{z})$ which constituted the basis of Borel's new study of P. A. Smith's theorems, [3]. We touch this point very lightly and recommend Borel's original paper to the reader.

There is a classifying space for complex vector bundles with operators which we do not go into in these notes, [2, 11] . For the case of line bundles with operators it can be described as follows. First there is the regular representation of π on C^k where $k = \#(\pi)$. Form the n-fold direct sum of the regular representation with itself to obtain a representation of π on C^{nk}. This representation sends lines into lines and so induces an action $(\pi, CP(nk-1))$. Since a vector in a line is also carried into a vector in the image line there is naturally induced an action $(\pi, \eta) \to (\pi, CP(nk-1))$ on the universal bundle. If X is reasonably nice, then complex line bundles with operators over (π, X) are in 1 - 1 correspondence with the equivariant homotopy classes of equivariant maps of (π, X) into $(\pi, CP(nk-1))$, for n sufficiently large. We shall show that these equivariant homotopy classes are in 1 - 1 correspondence with the elements of $H^2(\pi; \mathcal{J})$.

In section 1 the basic object is a pair (π, G) consisting of a left action of π on a topological group G as a finite group. Line bundles with operators correspond to a trivial action of π on $G = U(1)$ (or $G = C^*$ in the holomorphic case). The added generality, while making some formulas a bit awkward and stretching out some computations, notably in section 4, does provide a unified background in which line bundles with operators, Atiyah real line bundles, and certain other cases we shall suggest, can all be embedded. After the initial discussion of (π, G) the theory goes along until the necessity for explicit computations arise. We are sparing in the use of examples, but we encourage the reader to work out the case of Atiyah real line bundles for himself. It is an interesting case and tractable because $\pi = Z_2$.

We assume some familiarity with transformation groups. We use X/π for the quotient space and $\nu : X \to X/\pi$ for the quotient map. We use π_x to denote the isotropy group at $x \in X$; that is, the subgroup of elements leaving x fixed. In connection with sheaves, we show just how far behind times we are by using $\Gamma(U,\mathcal{A})$ to denote continuous sections. Otherwise we try to follow Glen Bredon's excellent treatment.

To sum up, we have evolved a quasi-algebraic approach parallel to Bredon's treatment of the original ideas in the study of transformation groups introduced by Borel and Cartan. We have done this to study the classification of complex line bundles with operators. The problem was set up and so analysed to illustrate how concepts and results from several branches of modern topology can be fused together within the study of transformation groups. There is nothing at all original in Chapter I. The theory of bundles with operators is long and this author is not really well acquainted with it. The first formal treatment was due to Palais. The theory has gone through several stages with the present emphasis on equivariant K-Theory.

In Chapter II we illustrate the use of bordism techniques in the study of periodic maps, in this case, of orientation preserving involutions on closed oriented manifolds. This matter was by-passed entirely until quite recently. The analogy between orientation preserving involutions on the one hand and involutions without regard to orientation, or orientation preserving maps of odd period, is quite tenuous and it was clear that some new thoughts would be appropriate.

Rosenzweig, [12], started the work on $\mathcal{O}_*(Z_2)$, the bordism algebra of all orientation preserving involutions on closed oriented manifolds. The first question is the relation to $\Omega_*(Z_2)$, the bordism module of fixed point free orientation preserving involutions on closed oriented manifolds. There is the natural homomorphism

$$i_* : \Omega_*(Z_2) \to \mathcal{O}_*(Z_2)$$

which ignores the freeness. The first key step was Rosenzweig's proof that the kernel of i_* consists of precisely all the 2-torsion in $\Omega_*(Z_2)$. One of the topics discussed in Chapter II is the image of i_*. Next there is introduced the bordism algebra $\mathcal{A}_*$. The basic object is an orientation preserving involution (T,B^n) on a compact oriented manifold with no fixed points in ∂B^n. Then (T,B^n) bounds if and only if there is an orientation preserving (τ, W^{n+1}) for which $(T,B^n) \subset (\tau, \partial W^{n+1})$ as a compact regular invariant submanifold and such that τ has no fixed points in $\partial W^{n+1} \setminus B^n$. Then $\mathcal{A}_n$ is the resulting bordism group. The triangle

$$\begin{array}{ccc} \mathcal{O}_*(Z_2) & \xrightarrow{j_*} & \mathcal{A} \\ & {\scriptstyle i_*}\nwarrow \quad \swarrow{\scriptstyle \partial_*} & \\ & \Omega_*(Z_2) & \end{array}$$

where j_* and ∂_* are defined in the obvious manner, is exact. Next Rosenzweig computes the structure of $\mathcal{A}$ as an Ω-module. The answer is not analogous to that of the case of maps of odd period, [7] .

We begin in section 1 of Chapter II with $\mathcal{A}$ again, but in the following format. The basic object is a pair $(\xi \to B^n, \sigma)$

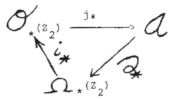

wherein $\xi \to B^n$ is an orthogonal 2k-plane bundle over a compact manifold together with an orientation $\mathcal{O}$ on the Whitney sum $\xi \oplus \tau \to B^n$ where $\tau \to B^n$ is the tangent n-plane bundle. Let $-(\xi \to B^n, \mathcal{O})$ be $(\xi \to B^n, -\mathcal{O})$, a boundary operator

$$\partial(\xi \to B^n, \mathcal{O}) = (\xi \to \partial B^n, \partial_*(\mathcal{O}))$$

by remaking that along ∂B^n, $\xi \oplus \tau = (\xi \oplus \tau') \oplus \eta$ where η is the outward pointing normal at each point of ∂B^n and then orienting $\xi \oplus \tau' \to \partial B^n$ compatibly with η and $\xi \oplus \tau$. The bordism group $A_n(2k)$ of such objects $(\xi \to M^n, \mathcal{O})$ wherein M^n is closed is then defined. It is not difficult to identify $A_n(2k)$ with the Atiyah bordism group $A_n(T, BSO(2k))$ where T is the covering involution over BO(2k). Then we are able to use Rosenzweig's result to see

$$A_n(2k) \cong \tilde{\Omega}_{n+1}(M(\gamma))$$

where $M(\gamma)$ is the Thom space of the canonical real line bundle $\gamma \to BO(2k)$.

Next we define a self-intersection homomorphism $S : A_n(2k) \to \Omega_{n-2k}(BO(k))$. This has been studied by others, including Hatori, but in different contexts. Roughly speaking take a generic cross-section $\chi: M^n \to \xi$ and intersect the image of χ with the o-section to obtain a closed regular submanifold $V^{n-2k} \subset M^n$. Now the normal bundle $\hat{\xi} \to V^{n-2k}$ is the restriction of ξ, therefore $\xi \oplus \tau | V^{n-2k}$ is

$\hat{\xi} \oplus (\hat{\xi} \oplus \hat{\tau}) = (\hat{\xi} \oplus \hat{\xi}) \oplus \hat{\tau}$. Now $\xi \oplus \tau$ is oriented and $\hat{\xi} \oplus \hat{\xi}$ is oriented by the complex structure $(v,w) \to (-w,v)$, so the tangent bundle $\hat{\tau} \to V^{n-2k}$ receives a compatible orientation. We set

$$S[\xi \to M^n, \theta] = [\hat{\xi} \to V^{n-2k}] \in \Omega_{n-2k}(BO(2k)).$$

We use a different but equivalent definition of self-intersection to see it is a function of bordism classes in $A_n(BO(2k))$. We set

$$\mathcal{a}_m = \sum_{n+2k=m} A_n(BO(2k))$$

$$\mathcal{M}_m = \sum_{q+4p=m} \Omega_q(BO(2p))$$

and $\mathcal{a} = \sum_0^\infty \mathcal{a}_m$, $\mathcal{M} = \sum_0^\infty \mathcal{M}_m$. Via the external Whitney sum both $\mathcal{a}$ and $\mathcal{M}$ are made into graded commutative algebras over Ω and $S : \mathcal{a} \to \mathcal{M}$ is an algebra homomorphism.

We use S to prove $\mathcal{a}/\mathrm{Tor}(\mathcal{a})$ is the graded polynomial ring over Ω/Tor generated by $[\eta_{2p+1} \to CP(2p+1), \theta]$, $p \geq 0$, where $\eta_{2p+1} \to CP(2p+1)$ is the Hopf bundle and $\eta_{2p+1} \oplus \tau \to CP(2p+1)$ is oriented by the complex structure. This is done by showing

$$S[\eta_{2p+1} \to CP(2p+1), \theta]$$
$$= -[\eta_{2p} \to CP(2p)]$$

and then by a standard we prove $\mathcal{M}/\mathrm{Tor}(\mathcal{M})$ is the polynomial algebra over Ω/Tor generated by $\{[\eta_{2p} \to CP(2p)]\}_{p=0}^\infty$.

Thus $S : \mathcal{A}/\text{Tor} \to \mathcal{M}/\text{Tor}$ is onto, but with Rosenzweig's computation we know enough about the module structure of $\mathcal{A}/\text{Tor}$ to see that S has no kernel.

Let $J \subset \mathcal{O}_*(Z_2)$ be the ideal which is the kernel of $j_* : \mathcal{O}_*(Z_2) \to \mathcal{A}$. There is an isomorphism $\Omega/\text{Tor} \cong J$ given by $[M^{4p}] \to [Z_2, Z_2][M^{4p}]$, where $[Z_2, Z_2]$ denotes Z_2 acting on itself by translation. Then let $[A, CP(2p)] \in \mathcal{O}_{4p}(Z_2)$, $p > 0$ be given by $A[z_1, \ldots, z_{2p+1}] = [-z_1, z_2, \ldots, z_{2p+1}]$ then with the self-intersection we show $(\Omega_*(Z_2)/J)/\text{Tor}$ is the polynomial ring over Ω/Tor generated by $\{[A, CP(2p)]\}_{p=1}^{\infty}$. Thus we know J and we know $\Omega_*(Z_2)/J$ modulo torsion. In theory we could determine the 2 rank of the torsion in $\mathcal{O}_*(Z_2)$ if we could determine

$$J \cap 2\,\mathcal{O}_*(Z_2).$$

It turns out that $[Z_2, Z_2][M^{4p}]$ lies in $2\,\mathcal{O}_{4p}(Z_2)$ if and only if index $[M^{4p}] = 0 \bmod 2$. This is done by using special examples and S plays a key role again.

In section 5 we introduce $\text{Tr} : \mathcal{O}_*(Z_2) \to Z$, that is, given (T, M^{4p}) put in $H^{2p}(M^{4p}; R)$ the usual inner-product $(v, w) = \langle v \cup w, \sigma_{4p} \rangle \in R$ and $(T^* v, T^* w) = (v, w)$ since T preserves orientation. Split $H^{2p}(M^{4p}; R)$ into $V^{ev} \oplus V^{od}$, the ± 1 eigenvalues of T^*. Then

$$\text{Tr}\,[T, M^{4p}] = \text{index}\,(V^{ev}) - \text{index}\,(V^{od}).$$

This is a bordism invariant and defines a ring homomorphism

$Tr : \mathcal{O}_*(Z_2) \to Z$ which vanishes on J. Since we know the structure of $\mathcal{O}_*(Z_2)/J$ mod torsion we can verify again the formula for Tr in terms of the index of the self-intersection of the fixed point set with itself; a formula drawn by Hirzebruch from the Atiyah-Bott Fixed Point Theorem.

Since index $[M^{4p}] = Tr[T, M^{4p}]$ mod 2 we define

$$\lambda : \mathcal{O}_{4p}(Z_2) \to Z_2$$

by $\lambda[T, M^{4p}] = 1/2(\mathrm{ind}[M^{4p}] - Tr[T, M^{4p}])$. We then prove

$$\mathcal{O}_{4p}(Z_2) \xrightarrow{2} \mathcal{O}_{4p}(Z_2) \xrightarrow{r+\lambda} I_{4p}(Z_2) \oplus Z_2$$

is exact. Here $I_{4p}(Z_2)$ is the unoriented analog of $\mathcal{O}_{4p}(Z_2)$ and r is the forgetful homomorphism.

In section 6 we work out some examples involving the conjugation involutions on almost complex manifolds of dimension 4p.

Chapter I

Line Bundles With Operators

1. A co-ordinate system with operators

We are all familiar with the co-ordinate transformations $g_{j,i}$ in Steenrod's book, and we also know how these determine a co-ordinate bundle, [13, p. 14]. Suppose $(\pi, X, \mathcal{V})$ is a triple consisting of a left action of the finite group π on a topological space X together with an indexed covering, $\mathcal{V}$, of X by π-invariant open sets. Fix a left action (π, G) of on a topological group G as a group of automorphisms.

A co-ordinate system with operators is an assignment to each ordered pair of indices (i,j) of a map

$$r_{j,i} : \pi \times (V_j \cap V_i) \longrightarrow G$$

such that for $x \in V_k \cap V_j \cap V_i$ and $h_1, h_2 \in \pi$

$$r_{k,i}(h_1 h_2, x) = (h_2^{-1} r_{k,j}(h_1, h_2 x)) r_{j,i}(h_2, x).$$

Remember π acts as a group of automorphisms on G.

(1.1) Lemma: A co-ordinate system with operators is equivalently determined by assigning to each (i,j) a map

$$R_{j,i} : \pi \times (V_j \cap V_i) \longrightarrow G$$

such that on $V_k \cap V_j \cap V_i$

$$R_{k,i}(h_1h_2,x) = (R_{k,j}(h_1,h_2x))(h_1R_{j,i}(h_2,x)).$$

Let $r_{j,i}(h,x) = h^{-1}R_{j,i}(h,x)$ then

$$\begin{aligned} r_{k,i}(h_1h_2,x) &= h_2^{-1}(h_1^{-1}(R_{k,i}(h_1h_2,x)) \\ &= h_2^{-1}(h_1^{-1}(R_{k,j}(h_1,h_2x)\cdot h_1R_{j,i}(h_2,x)) \\ &= h_2^{-1}r_{k,j}(h_1,h_2x)r_{j,i}(h_2,x). \end{aligned}$$

Conversely, given the $r_{j,i}(h,x)$ introduce

$R_{j,i}(h,x) = hr_{j,i}(h,x)$.

In any case note that if $e \in \pi$ is the identity

$$r_{k,i}(e,x) = r_{k,j}(e,x)r_{j,i}(e,x)$$

on $V_k \cap V_j \cap V_i$ so that the $\{r_{j,i}(e,x)\}$ form a co-ordinate system in the usual sense.

(1.2) Lemma: For any pair (i,j)

$$r_{j,j}(h,x)r_{j,i}(e,x) = r_{j,i}(h,x) = (h^{-1}r_{j,i}(e,hx))r_{i,i}(h,x)$$

on $V_j \cap V_i$.

Apply the defining equations first to $h_1 = h$, $h_2 = e$ and then to $h_1 = e$, $h_2 = h$.

Following [13, p. 14] we propose now to construct an associated right principal (π,G)-space, (π,B,G) over (π,X). As usual, consider the disjoint union $\bigsqcup (i \times V_i \times G)$ and say that $(i,x,g) \sim (j,x',g')$ if and only if $x = x'$ and $g' = r_{j,i}(e,x)g$.

Denote by $((i,x,g))$ the resulting equivalence class and by B the space of all such equivalence classes. A right principal action (B,G) is given by $((i,x,g))\hat{g} = ((i,x,g\hat{g}))$. The left action of π on B is then given by

$$h((i,x,g)) = ((i,hx,hr_{i,i}(h,x)\cdot hg)).$$

We must show this is well defined. Note by (1.2) that

$$hr_{j,j}(h,x)\cdot hr_{j,i}(e,x) = r_{j,i}(e,hx)\cdot hr_{i,i}(h,x)$$

thus if $g' = r_{j,i}(e,x)g$ then

$$h(r_{j,j}(h,x)\cdot g') = hr_{j,j}(h,x)\cdot hr_{j,i}(e,x)\cdot hg$$

$= r_{j,i}(e,hx)\ h(r_{i,i}(h,x)\cdot g)$. Hence the left action is well defined. For $h \in \pi$, $b \in B$ and $g \in G$ we have

$$h(bg) = h(b)hg$$

according to the above definitions. The projection map $p : B \to X$ given by $p((i,x,g)) = x$ is π-equivariant.

Given (π,X) we must define the expression "right principal (π,G)-space over (π,X)." We must then show that every such arises from a co-ordinate system with operators, and finally we must know when two co-ordinate systems with operators give rise to the same principal (π,G)-space.

A right principal (π,G)-space (π,B,G) is a triple consisting of a right principal G-space (B,G) together with a left action of π on B such that for $h \in \pi$, $b \in B$ and $g \in G$

$$h(bg) = h(b)\cdot h(g).$$

A right principal (π,G)-space, $p : (\pi,B,G) \to (\pi,X)$ over (π,X) consists of

1) a right principal (π,G)-space, (π,B,G).
2) an open onto mapping $p : B \to X$ such that
   a) p is a π-equivariant map
   b) $p(b) = p(b')$ if and only if there is a $g \in G$ with $b' = bg$
   c) for each $x \in X$ there is an open π-invariant neighborhood V_x and a map $\chi : V_x \to B$ with $p\chi$ = identity.

Let us first show how we can find a suitable co-ordinate system with operators. Cover X with a collection of open π-invariant sets, each having a specificed cross-section. Over $V_j \in \mathcal{V}$ denote by $\chi_j : V_j \to B$ this section. For any $x \in V_i \cap V_j$ and $h \in \pi$ we see from the π-equivariance of the projection map $p : B \to X$ that

$$p\,\chi_i(x) = p(h^{-1}\chi_j(hx)) = x$$

therefore there is a unique element $r_{j,i}(h,x) \in G$ for which

$$\chi_i(x) = (h^{-1}\chi_j(hx))\cdot r_{j,i}(h,x).$$

Now if $x \in V_k \cap V_j \cap V_i$ then

$$\chi_i(x) = h_2^{-1}\chi_j(h_2x)\cdot r_{j,i}(h_2,x)$$

$$h_2^{-1}\mathcal{X}_j(h_2x) = h_2^{-1}h_1^{-1}\mathcal{X}_k(h_1h_2x)\cdot h_2^{-1}r_{k,j}(h_1,h_2x)$$

$$\mathcal{X}_i(x) = h_2^{-1}h_1^{-1}\mathcal{X}_k(x)\cdot r_{k,i}(h_1h_2,x)$$

hence we have the required identity

$$r_{k,i}(h_1h_2,x) = (h_2^{-1}r_{k,j}(h_1,h_2x))(r_{j,i}(h_2,x)).$$

Denote by $(\pi,B',G) \xrightarrow{p} (\pi,X)$ the right principal (π,G)-space over (π,X) constructed from the co-ordinate system with operators. Define

$$\Phi : B' \longrightarrow B$$

by $\Phi((i,x,g)) = \mathcal{X}_i(x)g$. Note that if $((j,x',g')) = ((i,x,g))$ then $x = x' \in V_j \cap V_i$ and $g' = r_{j,i}(e,x)g$, thus

$$\Phi((j,x',g')) = \mathcal{X}_j(x)r_{j,i}(e,x)g,$$

but by definition $\mathcal{X}_i(x) = \mathcal{X}_j(x)r_{j,i}(e,x)$, hence Φ is well defined. The reader may define Φ^{-1}. In addition Φ is both G and π-equivariant. The G-equivariance is trivial and $h((i,x,g)) = ((i,hx,hr_{i,i}(h,x)\cdot hg))$ so $\Phi(h((i,x,g))) = \mathcal{X}_i(hx)\cdot h(r_{i,i}(h,x)g)$, but by definition $h\mathcal{X}_i(x) = \mathcal{X}_i(hx)\cdot hr_{i,i}(h,x)$ hence $\Phi(h((i,x,g))) = h(\mathcal{X}_i(x)g) = h\Phi(((i,x,g)))$. Finally,

$$\begin{array}{ccc} (\pi,B',G) & \xrightarrow{\ \Phi\ } & (\pi,B,G) \\ & {\scriptstyle p'}\searrow \qquad \swarrow{\scriptstyle p} & \\ & (\pi,X) & \end{array}$$

is a commutative diagram. This means every right principal (π,G)-space over (π,X) can be obtained from a co-ordinate system with operators.

Suppose next that on $(\pi,X,\mathcal{V})$ we have two co-ordinate systems with operators. Thus we have two sets $\{r_{j,i}(h,x)\}$ and $\{r'_{j,i}(h,x)\}$. We shall say that these are equivalent if and only if there are maps

$$\lambda_i : V_i \longrightarrow G$$

such that for $x \in V_i \cap V_j$ and $h \in \pi$

$$r'_{j,i}(h,x)\lambda_i(x) = h^{-1}\lambda_j(hx)\cdot r_{j,i}(h,x).$$

Let $p : (\pi,B,G) \longrightarrow (\pi,X)$ and $p' : (\pi,B',G) \longrightarrow (\pi,X)$ be the right principal (π,G)-spaces constructed respectively from $\{r_{j,i}(h,x)\}$ and $\{r'_{j,i}(h,x)\}$. We shall introduce an equivalence

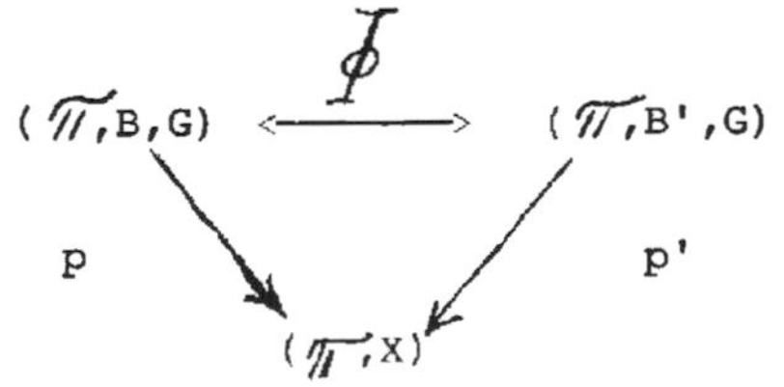

Let $\Phi(((i,x,g))) = ((i,x,\lambda_i(x)g))'$. The reader should verify that Φ is a well defined π-G, equivariant homeomorphism. Now observe the following

(1.3) Lemma: If for all i

$$r'_{i,i}(h,x)\lambda_i(x) = (h^{-1}\lambda_i(hx))r_{i,i}(h,x)$$

<u>and</u> <u>for</u> <u>all</u> (i,j)

$$r'_{j,i}(e,x)\lambda_i(x) = \lambda_j(x)r_{j,i}(e,x)$$

<u>then</u>

$$r'_{j,i}(h,x)\lambda_i(x) = (h^{-1}\lambda_j(hx))r_{j,i}(h,x).$$

Recall from (1.2) that

$$r_{j,j}(h,x)r_{j,i}(e,x) = r_{j,i}(h,x) = (h^{-1}r_{j,i}(e,hx))r_{i,i}(h,x)$$

and similarly for $r'_{j,i}(h,x)$. Thus

$$\begin{aligned}(h^{-1}\lambda_j(hx))r_{j,i}(h,x) &= h^{-1}(\lambda_j(hx)r_{j,i}(e,hx))\cdot r_{i,i}(h,x)\\ &= (h^{-1}(r'_{j,i}(e,hx)\lambda_i(hx)))r_{i,i}(h,x)\\ &= (h^{-1}r'_{j,i}(e,hx))(h^{-1}\lambda_i(hx)r_{i,i}(h,x))\\ &= (h^{-1}r'_{j,i}(e,hx))r'_{i,i}(h,x)\lambda_i(x)\\ &= r'_{j,i}(h,x)\lambda_i(x).\end{aligned}$$

Let $\mathcal{U}$ be a covering of X by π-invariant open sets which refines $\mathcal{V}$. That is, if J and J' are respectively the indexing sets of $\mathcal{V}$ and $\mathcal{U}$ there is a function $\tau: J' \longrightarrow J$ for which

$$U_{i'} \subset V_{\tau(i')}.$$

A co-ordinate system with operators $\{r_{j,i}(h,x)\}$ on $\mathcal{V}$ induces a co-ordinate system with operators on $\mathcal{U}$ by

$$r_{j',i'}(h,x) = r_{\tau(j'),\tau(i')}(h,x)$$

for $h \in \Pi$, $x \in U_{i'} \cap U_{j'}$. If $\tilde{\tau} : J' \longrightarrow J$ is a second refining function inducing

$$\tilde{r}_{j',i'}(h,x) = r_{\tilde{\tau}(j'),\tilde{\tau}(i')}(h,x)$$

then $\{r_{j',i'}(h,x)\}$ is equivalent to $\{\tilde{r}_{j',i'}(h,x)\}$. Since $U_{i'} \subset V_{\tau(i')} \cap V_{\tilde{\tau}(i')}$ we let $\lambda_i(x) = r_{\tau(i'),\tilde{\tau}(i')}(e,x) \in G$. On $U_{i'} \cap U_{j'}$ we have

$$r_{j',i'}(h,x) = r_{\tau(j'),\tau(i')}(h,x)$$

$$\tilde{r}_{j',i'}(h,x) = r_{\tilde{\tau}(j'),\tilde{\tau}(i')}(h,x).$$

Thus $r_{j',i'}(e,x)\lambda_{i'}(x) = r_{\tau(j'),\tilde{\tau}(i')}(e,x) = \lambda_{j'}(x)\tilde{r}_{j',i'}(e,x)$. Furthermore,

$$r_{i',i'}(h,x)\lambda_{i'}(x) = r_{\tau(i'),\tau(i')}(h,x)r_{\tau(i'),\tilde{\tau}(i')}(e,x)$$

$$= (h^{-1}r_{\tau(i'),\tilde{\tau}(i')}(e,hx))r_{\tilde{\tau}(i'),\tilde{\tau}(i')}(h,x)$$

$$= (h^{-1}\lambda_{i'}(hx))\tilde{r}_{i',i'}(h,x).$$

Thus if $\mathcal{U} < \mathcal{V}$ there is a unique natural transformation of the equivalence classes of co-ordinate systems with operators on $\mathcal{V}$ into those on $\mathcal{U}$. There is a direct limit set which is in natural 1 - 1 correspondence with the equivalence classes of the right principal (Π,G)-spaces.

A left (Π,G)-space F is a space on which Π and G both operate from the left so that if $h \in \Pi$, $g \in G$ any $y \in F$

then

$$h(g \cdot y) = h(g) \cdot hy.$$

To the right principal (π, G)-space $p : (\pi, B, G) \longrightarrow (\pi, X)$ we can associate a fibre bundle with fibre F and structure "system" (π, G). Thus let G act on $B \times F$ by $g(b,y) = (bg^{-1}, gy)$ and set $E = (B \times F)/G$, the quotient space of this left principal action. Denoting a point in E by $((b,y))$ the action (π, E) is given by

$$h((b,y)) = ((hb, hy)).$$

Since $h((bg^{-1}, gy)) = ((hb \cdot h(g^{-1}), h(g)hy))$ the action is well defined and the fibre map $(\pi, E) \longrightarrow (\pi, X)$ is π-equivariant.

Most attention has been given to the case where the action of π on G is trivial. On the other hand, Atiyah's K-Theory And Reality, [1] , corresponds to $G = U(n)$, $\pi = Z_2$ and acting on $U(n)$ as conjugation. Conjugation on C^n will make the affine space into a left $(Z_2, U(n))$-space and the associated bundles with fibre C^n receives a "conjugation."

The reader might ask if in our definition of right principal (π, G)-spaces there was a loss of generality in requiring cross-sections to exist over π-invariant open sets. The answer is no because at any $x \in X$ there are arbitrarily small open sets U_x invariant under the isotropy group π_x and which has the property that $hU_x \cap U_x \neq \emptyset$ if and only if $h \in \pi_x$. Given a section over such a U_x then by the action of π it is translated over $\bigcup_{h \in \pi} hU_x$.

2. Covariant stacks

Fix a simplicial complex K and associate to K the category $\mathcal{K}$ in which an object $\sigma \in \mathcal{K}$ is a non-empty set of vertices all belonging to some simplex of K and in which the morphisms are inclusions.

An elementary p-chain of K is an ordered (p+1)-tuple $c_p = (v^o,\ldots,v^p)$ of vertices all of which belong to some simplex of K. To the elementary p-chain c_p we associate $|c_p| \in \mathcal{K}$, the set of vertices $\{v^o,\ldots,v^p\}$. For each $\sigma \in \mathcal{K}$ denote by $C_p(\sigma)$ the free abelian group generated by all elementary p-chains, c_p, with $|c_p| \subset \sigma$. The usual face and degeneracy operators are

$$d_j : (v^o,\ldots,v^p) = (v^o,\ldots,\hat{v}^j,\ldots,v^p),\ o \leq j \leq p$$

$$s_i : (v^o,\ldots,v^p) = (v^o,\ldots,v^i,v^i,\ldots,v^p),\ o \leq i \leq p.$$

Thus $\{C_p(\sigma), d_j, s_i\} = C(\sigma)$ is a simplicial Z-module, If $\sigma' \subset \sigma$ then there is naturally induced a simplicial map $C(\sigma') \longrightarrow C(\sigma)$, hence $\sigma \longrightarrow C(\sigma)$ is a covariant functor from the category $\mathcal{K}$ into the category of simplicial Z-modules and maps.

A covariant stack, [14], $\mathcal{A} \rightarrow$ K is a covariant functor from $\mathcal{K}$ into the category of abelian groups and homomorphisms. We wish to define $H^p(K;\mathcal{A})$, thus we must first define the cochain group.

(2.1) Lemma: <u>The following definitions of the cochain</u>

group are equivalent

1) *A p-cochain* $\varphi \in C^p(K;\mathcal{A})$ *is a function which to each elementary p-chain,* c_p*, assigns an element* $\varphi(c_p) \in \mathcal{A}(|c_p|)$*. The coboundary* $\delta : C^p(K;\mathcal{A}) \to C^{p+1}(K;\mathcal{A})$ *is given by*

$$(\delta\varphi)(c_{p+1}) = \sum_{j=0}^{p+1} (-1)^j \mathcal{A}(|d_j c_{p+1}| \to |c_{p+1}|)\varphi(d_j(c_{p+1})).$$

2) *A p-cochain* $\Phi \in \mathcal{C}^p(K;\mathcal{A})$ *is a function which to each* $\sigma \in \mathcal{K}$ *assigns* $\Phi(\sigma) \in \mathrm{Hom}\,(C_p(\sigma), \mathcal{A}(\sigma))$ *so that if* $\sigma' \subset \sigma$ *the diagram*

$$\begin{array}{ccc} C_p(\sigma) & \xrightarrow{\Phi(\sigma)} & \mathcal{A}(\sigma) \\ \uparrow & & \uparrow \\ C_p(\sigma') & \xrightarrow{\Phi(\sigma')} & \mathcal{A}(\sigma') \end{array}$$

commutes. The coboundary $\delta : \mathcal{C}^p(K;\mathcal{A}) \to \mathcal{C}^{p+1}(K;\mathcal{A})$ *is given by taking* $(\delta\Phi)(\sigma)$ *to be the composite homomorphism*

$$C_{p+1}(\sigma) \xrightarrow{\partial} C_p(\sigma) \xrightarrow{\Phi(\sigma)} \mathcal{A}(\sigma).$$

The reader should verify that $\delta\Phi$ satisfies the requisite commutativity conditions. Define $F : C^p(K;\mathcal{A}) \to \mathcal{C}^p(K;\mathcal{A})$ as follows. Given $\varphi(c_p) \in \mathcal{A}(|c_p|)$ for each elementary p-chain, and an object $\sigma \in \mathcal{K}$, let $F(\varphi)(\sigma)$ be the homomorphism whose value on a c_p with $|c_p| \subset \sigma$ is $\mathcal{A}(|c_p| \to \sigma)\varphi(c_p)$. Then $G : \mathcal{C}^p(K;\mathcal{A}) \to C^p(K;\mathcal{A})$ is given by

$$(G(\Phi))(c_p) = \Phi(|c_p|)(c_p) \in \mathcal{A}(|c_p|).$$

We wish to show that FG = GF = identity. Now

$GF(\varphi)(c_p) = F(\varphi)(|c_p|)(c_p) = \varphi(c_p)$ and

$FG(\Phi)_\sigma(c_p) = \mathcal{Q}(|c_p| \to \sigma)(G(\Phi)(c_p))$

$= \mathcal{Q}(|c_p| \to \sigma)(\Phi(|c_p|))(c_p) = \Phi(\sigma)(c_p)$ by

the commutativity for all $c_p \in C_p(\sigma)$.

We shall just show that $\delta F = F\delta$. Now $(\delta\Phi)(\sigma)$ is that homomorphism which on $c_{p+1} \in C_{p+1}(\sigma)$ has the value

$\sum_{j=0}^{p+1} (-1)^j \Phi(\sigma)(d_j c_{p+1})$. However,

$$(\delta\varphi)(c_{p+1}) = \sum_{j=0}^{p+1} (-1)^j \mathcal{Q}(|d_j c_{p+1}| \to |c_{p+1}|)\varphi(d_j c_{p+1})$$

so by the functorial property

$$F(\delta\varphi)_\sigma(c_{p+1}) = \sum_{j=0}^{p+1} (-1)^j \mathcal{Q}(|d_j c_{p+1}| \to \sigma)\varphi(d_j c_{p+1})$$

$$= \sum_{j=0}^{p+1} (-1)^j F(\varphi)_\sigma(d_j c_{p+1}) = (\delta F(\varphi))_\sigma .$$

The relation $\delta G = G\delta$ is left to our reader. From either definition of the cochain complex we arrive at $H^p(K;\mathcal{Q})$.

3. Covariant stacks with operators

Let us open this section by recalling the un-normalized bar construction for the group Π, [10, p. 115]. First, $\beta_o(\Pi)$ is a free right $Z(\Pi)$-module on one generator, denoted by ( ). For $p > 0$, $\beta_p(\Pi)$ is a free right $Z(\Pi)$-module with a generator $(h_1,\ldots,h_p)$ for every ordered p-tuple of elements in Π. The degeneracy operators $s_i : \beta_p(\Pi) \to \beta_{p+1}(\Pi)$ are the $Z(\Pi)$-module homomorphisms given by

$$s_i(h_1,\ldots,h_p) = (h_1,\ldots,h_{i-1},e,h_i,\ldots,h_p),\ 0 \leq i \leq p.$$

The face operators $d_j : \beta_p(\Pi) \to \beta_{p-1}(\Pi)$, $0 \leq j \leq p$, are the $Z(\Pi)$-module homomorphisms given by

$$d_o(h_1,\ldots,h_p) = (h_2,\ldots,h_p)$$

$$d_j(h_1,\ldots,h_p) = (h_1,\ldots,h_j h_{j+1},\ldots,h_p),\ 0 < j < p$$

$$d_p(h_1,\ldots,h_p) = (h_1,\ldots,h_{p-1})h_p.$$

In particular, $d_o(h_1) = (\ )$, $d_1(h_1) = (\ )h_1$. We can also define an augmentation $\epsilon : \beta_o(\Pi) \to Z$ by $(\ ) \to 1$. Thus $\beta(\Pi) = \{\beta_p(\Pi), d_j, s_i\}_{p=0}^{\infty}$ is a simplicial $Z(\Pi)$-module.

A basic o-chain, b_o, in K is a pair $(V^o;(\ ))$ wherein V^o is a vertex of K. For $p > 0$, basic p-chain, b_p, is a pair $(V^o,\ldots,V^p;h_1,\ldots,h_p)$ wherein $(V^o,\ldots,V^p)$ is an elementary p-chain. Set $|b_p| \in \mathcal{K}$ equal to the set $\{V^o,\ldots,V^p\}$. Denote by $B_p(\sigma)$ the free right $Z(\Pi)$-module generated by all the basic p-chains, b_p, with $|b_p| \subset \sigma$. Clearly $B_p(\sigma) \simeq C_p(\sigma) \otimes_Z \beta_p(\pi)$ and face and degeneracy operators can be directly defined in $B_p(\sigma)$ so that there is a natural simplicial $Z(\sigma)$-module isomorphism

$$\{B_p(\sigma), d_j, s_i\}_{p=0}^{\infty} \simeq \{C_p(\sigma) \otimes_Z \beta_p(\pi), d_j \otimes d_j, s_i \otimes s_i\}_{p=0}^{\infty} = C(\sigma) \times \beta(\pi).$$

A covariant stack with operators $\mathcal{A} \rightarrow K$ is a covariant functor from $\mathcal{K}$ to the category of right $Z(\pi)$-modules and homomorphisms. We wish to define $H^p(\pi;\mathcal{A})$, the cohomology of the group π with coefficients in a covariant stack having operators. This time there are four equivalent definitions of the p-cochains.

1) A p-cochain is a function φ which to each basic p-chain, b_p, assigns an element $\varphi(b_p) \in \mathcal{A}(|b_p|)$.

2) A p-cochain is a function Φ which to each $\sigma \in \mathcal{K}$ assigns an element $\Phi(\sigma)$ in $\mathrm{Hom}_{Z(\pi)}(B_p(\sigma), \mathcal{A}(\sigma))$ so that if $\sigma' \subset \sigma$ then the diagram

$$\begin{array}{ccc} B_p(\sigma) & \xrightarrow{\Phi(\sigma)} & \mathcal{A}(\sigma) \\ \uparrow & & \uparrow \\ B_p(\sigma') & \xrightarrow{\Phi(\sigma')} & \mathcal{A}(\sigma') \end{array}$$

commutes.

3) A p-cochain is a function Φ which to each $\sigma \in \mathcal{K}$ assigns $\Phi(\sigma) \in \mathrm{Hom}_Z(C_p(\sigma), \mathrm{Hom}_{Z(\pi)}(\beta_p(\pi), \mathcal{A}(\sigma)))$ so that if $\sigma' \subset \sigma$ the diagram

$$\begin{array}{ccc} C_p(\sigma) & \xrightarrow{\Phi(\sigma)} & \mathrm{Hom}_{Z(\pi)}(\mathcal{B}_p(\pi), \mathcal{A}(\sigma)) \\ \uparrow & & \uparrow \\ C_p(\sigma') & \xrightarrow{\Phi(\sigma')} & \mathrm{Hom}_{Z(\pi)}(\mathcal{B}_p(\pi), \mathcal{A}(\sigma')) \end{array}$$

commutes.

4) A p-cochain is a function φ which to each elementary p-chain c_p assigns $\varphi(c_p) \in \mathrm{Hom}_{Z(\pi)}(\mathcal{B}_p(\pi), \mathcal{A}(|c_p|))$.

The first and second, as well as the third and fourth, are seen to be equivalent by an argument similar to (2.1) If we recall that $B_p(\sigma) \simeq C_p(\sigma) \otimes_Z \mathcal{B}_p(\pi)$ then the standard isomorphism

$$\mathrm{Hom}_{Z(\pi)}(C_p(\sigma) \otimes_Z \mathcal{B}_p(\pi), \mathcal{A}(\sigma))$$
$$\updownarrow$$
$$\mathrm{Hom}_Z(C_p(\sigma), \mathrm{Hom}_{Z(\pi)}(\mathcal{B}_p(\pi), \mathcal{A}(\sigma)))$$

is used to establish the equivalence of the second and third definitions. If we omit, as is customary, the homomorphisms, then the coboundary operator in the first definition is given by

$$\delta\varphi(v^o, \ldots, v^{p+1}; h_1, \ldots, h_{p+1})$$
$$= \varphi(v^1, \ldots, v^{p+1}; h_2, \ldots, h_{p+1}) + \sum_{j=1}^{p} (-1)^j \varphi(v^o, \ldots, \hat{v}^j, \ldots, v^{p+1};$$
$$h_1, \ldots, h_j h_{j+1}, \ldots, h_{p+1}) + (-1)^{p+1} \varphi(v^o, \ldots, v^p; h_1, \ldots, h_p) h_{p+1}.$$

In any case we arrive at $H^p(\pi; \mathcal{A})$.

4. An example

In complex k-space, C^k, let $Z^k \subset C^k$ be the additive subgroup of points all of whose co-ordinates are integers. There is an exact sequence

$$0 \longrightarrow Z^k \longrightarrow C^k \xrightarrow{e} G \longrightarrow 0$$

where G is the k-fold direct product of C^*, the multiplicative abelian group of non-zero complex numbers, with itself, and $e(z_1,\ldots,z_k) = (\exp 2\pi i\, z_1,\ldots,\text{ex}\ 2\pi i\, z_k)$. We fix a left action of π on C^k as a group of complex linear automorphisms, and we assume Z^k is invariant under this action of π. There is naturally induced a left action of π on G as a group of automorphisms.

Consider then a left action (π, V^n) on a closed connected compact analytic manifold as a group of holomorphic transformations. Fix a covering $\mathcal{V}$ of V^n by open π-invariant subsets. Let K be the nerve of this covering. Since a vertex of K is an element of $\mathcal{V}$ to each $\sigma \in \mathcal{K}$ we can associate an open, non-empty π-invariant set, Sup $(\sigma) \subset V^n$ which is simply the intersection of the vertices which make up σ. A covariant stack with operators $\mathcal{G} \to$ K is defined as follows. For each $\sigma \in \mathcal{K}$, let $\mathcal{G}(\sigma)$ be the multiplicative abelian group of all holomorphic maps f : Sup $(\sigma) \to$ G. This is covariant for if $\sigma' \subset \sigma$ then Sup $(\sigma) \subset$ Sup (σ') and $\mathcal{G}(\sigma' \to \sigma)$ is the restriction. A right $Z(\pi)$-module structure on $\mathcal{G}(\sigma)$ is given by

$$((f)h)(x) = h^{-1}f(hx).$$

If $\varphi \in C^1(\pi;\mathcal{G})$ then to each $(V_j, V_i; h)$ there is associated a holomorphic function $r_{j,i}(h,x) : V_j \cap V_i \to G$. Now

$$(\delta\varphi)(V_k, V_j, V_i; h_1, h_2) = \varphi(V_j, V_i; h_2)\,\varphi(V_k, V_i; h_1, h_2)^{-1} \times \varphi(V_k, V_j; h_1)h_2,$$

which corresponds to

$$r_{j,i}(h_2,x)\,r_{k,i}(h_1h_2,x)^{-1}h_2^{-1}r_{k,j}(h_1,h_2x) .$$

Thus φ is a 1 - cocycle if and only if $\{r_{j,i}(h,x)\}$ is a holomorphic co-ordinate system with operators. Suppose that φ, φ' are two 1 - cocycles with $\{r_{j,i}(h,x)\}$, $\{r'_{j,i}(h,x)\}$ the corresponding co-ordinate systems with operators. If $\psi \in C^0(\pi;\mathcal{G})$ has $\varphi\delta\psi = \varphi'$ then ψ is an assignment to each V_i of a holomorphic $\lambda_i(x) : V_i \to G$ such that

$$r_{j,i}(h,x)\,\lambda_i(x)\,(h^{-1}\lambda_j(hx))^{-1} = r'_{j,i}(h,x)$$

or

$$r_{j,i}(h,x)\cdot\lambda_i(x) = (h^{-1}\lambda_j(hx))\,r'_{j,i}(h,x)$$

on $V_j \cap V_i$. Thus φ' and φ are cohomologous if and only if the corresponding co-ordinate systems with operators are holomorphically equivalent. Thus $H^1(\pi;\mathcal{G})$ is the abelian group of holomorphic equivalence classes of the holomorphic co-ordinate systems with operators over $\mathcal{V}$. Similarly, $H^1(K;\mathcal{G})$ is the group of holomorphic co-ordinate systems over $\mathcal{V}$. Let us study the relation of these two cohomology groups.

There is the natural homomorphism $H^1(\pi;\mathcal{G}) \to H^1(K;\mathcal{G})$ which ignores the operators. Consider a holomorphic co-ordinate system with operators $\{r_{j,i}(h,x)\}$ which represents an element of the kernel of this homomorphism. By definition, then, for each V_i there is a holomorphic $\lambda_i(x) : V_i \to G$ such that on $V_j \cap V_i$

$$r_{j,i}(e,x)\,\lambda_i(x) = \lambda_j(x)$$

We assert that on $V_j \cap V_i$

$$h^{-1}\lambda_j(hx)^{-1}\cdot r_{j,j}(h,x)\,\lambda_j(x) \equiv h^{-1}\lambda_i(hx)^{-1}\cdot r_{i,i}(h,x)\lambda_i(x).$$

Note first that on $V_j \cap V_i$

$$r_{j,j}(h,x) = r_{j,j}(h,x)r_{j,i}(e,x)r_{j,i}(e,x)^{-1}$$

$$= (h^{-1}r_{j,i}(e,hx))r_{i,i}(e,x)r_{j,i}(e,x)^{-1},$$

then

$$h^{-1}\lambda_j(hx)^{-1}\cdot r_{j,j}(h,x)\cdot\lambda_j(x)$$

$$= h^{-1}(\lambda_j(hx)^{-1}r_{j,i}(e,hx))\cdot(r_{i,i}(h,x)r_{j,i}(e,x)^{-1}\lambda_j(x))$$

$$= h^{-1}\lambda_i(hx)^{-1}\cdot r_{i,i}(h,x)\cdot\lambda_i(x).$$

This defines, therefore, for each $h \in \pi$ a unique holomorphic map $R(h,x) : V^n \to G$. Since V^n is closed and connected we can simply write $R : \pi \to G$. We want to see that $R(h_1h_2) = h_2^{-1}R(h_1)\cdot R(h_2)$. On V_j we write

$$h_2^{-1}h_1^{-1}\lambda_j(h_1h_2x)^{-1}\cdot r_{j,j}(h_1h_2,x)\cdot\lambda_j(x)$$

$$= h_2^{-1}h_1^{-1}\lambda_j(h_1h_2x)^{-1}\cdot h_2^{-1}r_{j,j}(h_1,h_2x)\cdot r_{j,j}(h_2,x)\lambda_j(x)$$

$$= h_2^{-1}(h_1^{-1}(\lambda_j(h_1h_2x)^{-1})\cdot r_{j,j}(h_1,h_2x)\lambda_j(h_2x))$$

$$\times(h_2^{-1}(\lambda_j(h_2x)^{-1})\circ r_{j,j}(h_2,x)\lambda_j(x)) = h_2^{-1}R(h_1)\cdot R(h_2).$$

If G is given the right $Z(\pi)$-module structure $gh = h^{-1}(g)$ then R defines a cohomology class in $H^1(\pi;G)$. By (1.3) we see $\{r_{j,i}(h,x)\}$ is equivalent to $R_{j,i}(h,x) = R(h)$. That is, let $\mu_j(x) = 1/\lambda_j(x)$ then

$$R(h)\mu_j(x) = h^{-1}\mu_j(hx)\cdot r_{j,j}(h,x)$$

$$\mu_i(x) = \mu_j(x)r_{j,i}(e,x) .$$

(4.1) Lemma: There is an exact sequence

$$0 \to H^1(\pi;G) \to H^1(\pi;\mathcal{G}) \to H^1(K;\mathcal{G}).$$

Now $H^1(K;\mathcal{G})$ is acted on from the right by π as a group of automorphisms; that is, if $\{g_{j,i}(x)\}$ is a holomorphic coordinate system, then so is $\{g^h_{j,i}(x) = h^{-1}g_{j,i}(hx)\}$, for

$$g^h_{k,i}(x) = h^{-1}g_{k,i}(hx) = h^{-1}(g_{k,j}(hx)g_{j,i}(hx)) = g^h_{k,j}(x)\cdot g^h_{j,i}(x)$$

Let $H^1(K;\mathcal{G})^{\pi}$ be the subgroup of elements fixed under the action of every $h \in \pi$. Let us prove that the image of $H^1(\pi;\mathcal{G}) \to H^1(K;\mathcal{G})$ lies in $H^1(K;\mathcal{G})^{\pi}$. For each $h \in \pi$ and each V_i, let $\lambda_i(h,x) = r_{i,i}(h,x) : V_i \to G$. Since

$r_{j,j}(h,x)r_{j,j}(e,x) = h^{-1}r_{j,i}(e,hx)\cdot r_{i,i}(h,x)$ on $V_i \cap V_j$ we have

$$h^{-1}r_{j,i}(e,hx)\cdot\lambda_i(h,x) = \lambda_j(h,x)\cdot r_{j,i}(e,x)$$

hence $\{h^{-1}r_{j,i}(e,hx)\}$ and $\{r_{j,i}(e,x)\}$ are equivalent co-ordinate systems.

This suggests a homomorphism $H^1(K;\mathcal{G})^{\pi} \longrightarrow H^2(\pi;G)$. Consider a $\{g_{j,i}(x)\}$ and suppose that for each $h \in \pi$ there are holomorphic $\lambda_i(h,x) : V_i \longrightarrow G$ with

$$h^{-1}g_{j,i}(hx)\cdot\lambda_i(h,x) = \lambda_j(h,x)g_{j,i}(x).$$

In particular, we must have

$$h_1^{-1}g_{j,i}(h_1h_2x)\cdot\lambda_i(h_1,h_2x) = g_{j,i}(h_2x)\lambda_j(h_1,h_2x)$$

$$h_2^{-1}g_{j,i}(h_2x)\cdot\lambda_i(h_2,x) = g_{j,i}(x)\lambda_j(h_2,x)$$

which combines to yield

$$h_2^{-1}(h_1^{-1}g_{j,i}(h_1h_2x)\cdot\lambda_i(h_1,h_2x)\cdot h_2\lambda_i(h_2,x))$$
$$= g_{j,i}(x)h_2^{-1}\lambda_j(h_1,h_2x)\cdot\lambda_j(h_2,x).$$

Now $h_2^{-1}h_1^{-1}g_{j,i}(h,h_2x)\cdot\lambda_i(h_1h_2,x) = g_{j,i}(x)\lambda_j(h_1h_2,x)$ also, thus

$$g_{j,i}(x)h_2^{-1}\lambda_j(h_1,h_2x)\cdot\lambda_j(h_2,x)$$
$$= h_2^{-1}h_1^{-1}g_{j,i}(h_1h_2x)\cdot h_2^{-1}\lambda_i(h_1,h_2x)\cdot\lambda_i(h_2,x)$$
$$= g_{j,i}(x)\lambda_j(h_1h_2,x)\cdot\lambda_i(h_1h_2,x)^{-1}\cdot h_2^{-1}\lambda_i(h_1,h_2x)\lambda_i(h_2,x).$$

We can now assert that on $V_j \cap V_i$

$$h_2^{-1}\lambda_j(h_1,h_2x)\cdot\lambda_j(h_2,x)\cdot\lambda_j(h_1h_2,x)^{-1}$$

$$h_2^{-1}\lambda_i(h_1,h_2x)\cdot\lambda_i(h_2,x)\cdot\lambda_i(h_1h_2,x)^{-1}.$$

Since V^n is closed and connected we can define a 2 -cochain $\varphi \in C^2(\pi;G)$ by assigning to the pair (h_1,h_2) the element

$$h_2^{-1}\lambda_j(h_1,h_2x)\cdot\lambda_j(h_2,x)\cdot\lambda_j(h_1h_2,x)^{-1},$$

which is independent of x and j. We could define $\psi \in C^1(\pi;\mathcal{G})$ by $\psi(V_j,V_i;h) = \lambda_j(h,x)$, then $\delta\psi(V_j,V_j,V_j;h_1h_2) = \lambda_j(h_2,x)\cdot\lambda_j(h_1h_2,x)^{-1}\cdot h_2^{-1}\lambda_j(h_1,h_2x)$. This is the easiest way to see that $\varphi \in Z^2(\pi;G)$. It is rather tedious to verify that this defines a homomorphism $H^1(K;\mathcal{G})^{\pi} \longrightarrow H^2(\pi;G)$.

(4.2) <u>Theorem</u>: <u>There is an exact sequence</u>

$$0 \longrightarrow H^1(\pi;G) \longrightarrow H^1(\pi;\mathcal{G}) \longrightarrow H^1(K;\mathcal{G})^{\pi} \longrightarrow H^2(\pi;G).$$

Suppose there is a function $\psi:\pi \longrightarrow G$ with

$$\psi(h_2)\psi(h_1h_2)^{-1}h_2^{-1}\psi(h_1) = \lambda_j(h_2,x)\lambda_j(h_1h_2,x)^{-1}h_2^{-1}\lambda_j(h_1,h_2x)$$

for all pairs (h_1,h_2). Then

$$\psi(h_1h_2)^{-1}\lambda_j(h_1h_2,x) = \psi(h_2)^{-1}\lambda_j(h_2,x)\cdot h_2^{-1}(\psi(h_1)\cdot\lambda_j(h_1,h_2x)).$$

Let us therefore introduce

$$r_{j,j}(h,x) = \psi(h)^{-1}\lambda_j(h,x)$$

then $r_{j,j}(h_1h_2,x) = h_2^{-1}r_{j,j}(h_1,h_2x)r_{j,j}(h_2,x)$ and in addition

$$\begin{aligned} r_{j,j}(h,x)g_{j,i}(x) &= \psi(h)^{-1}\lambda_j(h,x)g_{j,i}(x) \\ &= \psi(h)^{-1}\lambda_i(h,x)h^{-1}g_{j,i}(hx) \\ &= h^{-1}g_{j,i}(hx)\cdot r_{i,i}(h,x). \end{aligned}$$

Thus $r_{j,i}(h,x) = \psi(h)^{-1}\lambda_j(h,x)g_{j,i}(x)$ extends $\{g_{j,i}(x)\}$ to a holomorphic co-ordinate system with operators.

(4.3) <u>Corollary</u>: <u>If (π, V^n) has at least one stationary point then</u>

$$H^1(\pi;\mathcal{G}) \simeq H^1(\pi;G) \oplus H^1(K;\mathcal{G})^{\pi}.$$

We must first show that $H^1(K;\mathcal{G})^{\pi} \longrightarrow H^2(\pi;G)$ is trivial. Choose a stationary point p and suppose $p \in V_k$. Let us note that

$$\lambda_k(h_2,p)\lambda_k(h_1h_2,p)^{-1}h_2^{-1}\lambda_k(h_1,h_2p) = \lambda_k(h_2,p)\lambda_k(h_1h_2,p)^{-1}h_2^{-1}\lambda_k(h_1,p)$$

for all (h_1,h_2). Thus if $\psi(h) = \lambda_k(h,p)$, we have

$$\begin{aligned} (\delta\psi)(h_1,h_2) &= \lambda_k(h_2,p)\cdot\lambda_k(h_1,h_2,p)^{-1}\cdot h_2^{-1}\lambda_k(h_1,h_2p) \\ &= \lambda_j(h_2,x)\cdot\lambda_j(h_1,h_2,x)^{-1}\cdot h_2^{-1}\lambda_j(h_1,h_2x). \end{aligned}$$

The splitting homomorphism $H^1(\pi;\mathcal{G}) \longrightarrow H^1(\pi;G)$ is described by setting $R(h) = r_{k,k}(h,p)$ so

$$R(h_1h_2) = r_{k,k}(h_1h_2,p) = h_2^{-1}r_{k,k}(h_1,p)\cdot r_{k,k}(h_2,p) = h_2^{-1}R(h_1)R_2(h_2).$$

In the presence of a stationary point it would seem that the computation of $H^1(\pi;\mathcal{G})$ is more tractable. We can use the exact coefficient sequence of right $Z(\pi)$-modules

$$0 \longrightarrow Z^k \longrightarrow C^k \xrightarrow{e} G \longrightarrow 0$$

to show $H^1(\pi;G) \simeq H^2(\pi;Z^k)$ since $H^p(\pi;C^k) = 0$, $p > 0$. What we hope to develope later is a way of coming to terms with $H^1(K;\mathcal{G})^\pi$.

5. Two spectral sequences

We noted that $\sigma \rightarrow C(\sigma) \times \mathcal{B}(\pi)$ is a functor from $\mathcal{K}$ into the category of right simplicial $Z(\pi)$-modules and maps. Certainly we should introduce a second functor $\sigma \rightarrow C(\sigma) \otimes_Z \mathcal{B}(\pi)$ on $\mathcal{K}$ to the category of right $Z(\pi)$ chain complexes and maps by setting

$$(C(\sigma) \otimes \mathcal{B}(\pi))_p = \sum_{r+s=p} C_r(\sigma) \otimes \mathcal{B}_s(\pi).$$

As usual, $\partial_1 : C_r(\sigma) \otimes \mathcal{B}_s(\pi) \rightarrow C_{r-1}(\sigma) \otimes \mathcal{B}_s(\pi)$ is $\partial \otimes 1$ while $\partial_2 : C_r(\sigma) \otimes \mathcal{B}_s(\pi) \rightarrow C_r(\sigma) \otimes \mathcal{B}_{s-1}(\pi)$ is $(-1)^r \otimes \partial$, so that $\partial = \partial_1 + \partial_2$ is the boundary operator of the bigraded complex $C(\sigma) \otimes \mathcal{B}(\pi)$.

The Eilenberg-Zilberg Theorem [10, p 239] asserts that for each $\sigma \in \mathcal{K}$ there is a natural chain equivalence

$$C(\sigma) \times \mathcal{B}(\pi) \underset{g_\sigma}{\overset{f_\sigma}{\rightleftarrows}} C(\sigma) \otimes \mathcal{B}(\pi)$$

such that $f_\sigma g_\sigma$ = identity and $g_\sigma f_\sigma$ is chain homotopic to the identity by a natural chain homotopy. The naturality implies f_σ, g_σ and the homotopies are all $Z(\pi)$-module homomorphisms. We may, therefore, just as well use $\sigma \rightarrow C(\sigma) \otimes \mathcal{B}(\pi)$ to define $H^p(\pi;\mathcal{A})$. Thus $\phi \in C^{r,s}(\pi;\mathcal{A})$ is a function which to each $\sigma \in \mathcal{K}$ assigns $\phi(\sigma)$ in $\mathrm{Hom}_{Z(\pi)}(C_r(\sigma) \otimes \mathcal{B}_s(\pi), \mathcal{A}(\sigma))$ such that if $\sigma' \subset \sigma$ the diagram

$$\begin{array}{ccc} C_r(\sigma)\otimes\mathcal{B}_s(\pi) & \xrightarrow{\Phi(\sigma)} & \mathcal{Q}(\sigma) \\ \uparrow & & \uparrow \\ C_r(\sigma')\otimes\mathcal{B}_s(\pi) & \xrightarrow{\Phi(\sigma')} & \mathcal{Q}(\sigma') \end{array}$$

commutes. Let $(\delta''\Phi)(\sigma)$ be the composite homomorphism

$$C_r(\sigma)\otimes\mathcal{B}_{s+1}(\pi) \xrightarrow{\partial_2} C_r(\sigma)\otimes\mathcal{B}_s(\pi) \xrightarrow{\Phi(\sigma)} \mathcal{Q}(\sigma),$$

while $(\delta'\Phi)(\sigma)$ is the composite

$$C_{r+1}(\sigma)\otimes\mathcal{B}_s(\pi) \xrightarrow{\partial_1} C_r(\sigma)\otimes\mathcal{B}_s(\pi) \xrightarrow{\Phi(\sigma)} \mathcal{Q}(\sigma) .$$

With $\delta = \delta' + \delta''$ the Eilenberg-Zilber Theorem provides the cochain equivalence between this definition of $C^p(\pi;\mathcal{Q})$ and the second given in section 3. Since $(\delta')^2 = (\delta'')^2 = 0 = \delta'\delta'' + \delta''\delta'$ we obtain two spectral sequences for $H^*(\pi;\mathcal{Q})$, [5, p. 255 or 10].

A direct description of $C^{r,s}(\pi;\mathcal{Q})$ is given by noting $\varphi \in C^{r,s}(\pi;\mathcal{Q})$ is a function which to each elementary r-chain, c_r, assigns $\varphi(c_r) \in \mathrm{Hom}_{Z(\pi)}(\mathcal{B}_s(\pi), \mathcal{Q}(|c_r|))$. Now $\delta''\varphi(c_r) \in C^{r,s+1}(\pi;\mathcal{Q})$ is the composition $\mathcal{B}_{s+1}(\pi) \xrightarrow{(-1)^r\partial} \mathcal{B}_s(\pi) \xrightarrow{\varphi(c_r)} \mathcal{Q}(|c_r|)$ and $\delta'\varphi(c_{r+1})$ is the sum of the composite homomorphisms

$$\sum_{j=0}^{p+1} (-1)^j \mathcal{Q}(|d_j c_{r+1}| \to |c_{r+1}|)\varphi(d_j c_{r+1}).$$

Now

$${}''H^{r,s}(C(\pi;\mathcal{Q})) = \ker(\delta'': C^{r,s}(\pi;\mathcal{Q}) \to C^{r,s+1}(\pi;\mathcal{Q}))/\delta'' C^{r,s-1}(\pi;\mathcal{Q}).$$

Thus $\psi \in {}''H^{r,s}(C(\pi;\mathcal{Q}))$ is a function which to each elementary r-chain assigns an element $\psi(c_r) \in H^s(\pi;\mathcal{Q}(|c_r|))$. Let, therefore, for $s \geq 0$, $\mathcal{h}^s \longrightarrow K$ be the covariant stack which to each $\sigma \in \mathcal{K}$ assigns $H^s(\pi;\mathcal{Q}(\sigma))$, so that

$$'H^r({}''H^s(C(\pi;\mathcal{Q})) = H^r(K;h^s).$$

(5.1) Theorem: If $\mathcal{Q} \rightarrow K$ is a covariant stack with operators then there is a cohomology spectral sequence $\{'E_t^{r,s}, d_t\} \Rightarrow H^*(\pi;\mathcal{Q})$ with

$$'E_2^{r,s} \simeq H^r(K;\mathcal{h}^s).$$

Now we consider the other spectral sequence. Note that for a covariant stack with operators both $C^p(K;\mathcal{Q})$ and $H^p(K;\mathcal{Q})$ are right $Z(\pi)$-modules. We can write $C^{r,s}(\pi;\mathcal{Q}) = \mathrm{Hom}_{Z(\pi)}(\beta_s(\pi), C^r(K;\mathcal{Q}))$. This is clear since $\beta_s(\pi)$ is a free $Z(\pi)$-module; that is, $\bar{\phi} \in \mathrm{Hom}_{Z(\pi)}(\beta_s(\pi), C^r(K;\mathcal{Q}))$ is a function which to each generator H of $\beta_s(\pi)$ assigns $\bar{\phi}(H) \in C^r(K;\mathcal{Q})$, and thus $\bar{\phi}(H)(c_r) \in \mathcal{Q}(|c_r|)$ for every elementary r-chain. Now $(\delta'\bar{\phi})(H)$ is given by the coboundary operator $C^r(K;\mathcal{Q}) \longrightarrow C^{r+1}(K;\mathcal{Q})$ and

$$(-1)^r(\delta''\bar{\phi})(h_1,\ldots,h_{s+1})$$
$$= \bar{\phi}(h_2,\ldots,h_{s+1}) + \sum_{j=1}^{s} (-1)^j \bar{\phi}(h_1,\ldots,h_j h_{j+1},\ldots,h_{s+1})$$
$$+ (-1)^{s+1}\bar{\phi}(h_1,\ldots,h_s)h_{s+1}.$$

It follows immediately that

$$''H^s('H^r(C(\pi;\mathcal{A}))) \approx H^s(\pi;H^r(K;\mathcal{A})),$$

thus we have

(5.2) Theorem: If $\mathcal{A} \longrightarrow K$ is a covariant stack with operators, then there is a cohomology spectral sequence $\{''E_t^{s,r}, d_t\} \Rightarrow H^*(\pi;\mathcal{A})$ with

$$''E_2^{s,r} \simeq H^s(\pi;H^r(K;\mathcal{A})).$$

Suppose that each $\mathcal{A}(\sigma)$ is a free $Z(\pi)$-module, then $\mathcal{H}^s = 0$ for $s > 0$ and $\mathcal{H}^0 \longrightarrow K$ is the covariant stack $\mathcal{A}^{\pi} \longrightarrow K$ which to $\sigma \in \mathcal{K}$ assigns that subgroup of $\mathcal{A}(\sigma)$ consisting of all elements fixed under the action of π. In this case, $H^p(\pi;\mathcal{A}) \simeq H^p(K;\mathcal{A}^{\pi})$. As a corollary, then

(5.3) Corollary (Cartan): If $\mathcal{A} \longrightarrow K$ is a covariant stack with operators for which each $\mathcal{A}(\sigma)$ is a free $Z(\pi)$-module, then there is a spectral sequence $\{E_t^{s,r} d_t\} \Rightarrow H^*(K;\mathcal{A}^{\pi})$ for which

$$E_2^{s,r} \simeq H^s(\pi;H^r(K;\mathcal{A})).$$

This is precisely the analog of the cohomology spectral sequence of a covering.

6. Sheaves with operators

Fix a paracompact Hausdorff space. A sheaf with operators $p : (\mathcal{A},\pi) \to Y$ is a pair consisting of a sheaf together with a stalk preserving right action of π as a group of sheaf automorphisms. We regard the sheaves with operators together with their equivariant homomorphisms as forming a category.

For each open set $U \subset Y$ the continuous section $\Gamma(U,\mathcal{A})$ have a natural right $Z(\pi)$-module structure given by $(sh)(x) = s(x)\cdot h$. We denote by $\Lambda^s(\mathcal{A}) \to X$ the sheaf defined by the presheaf

$$U \to \operatorname{Hom}_{Z(\pi)}(\beta_s(\pi),\Gamma(U,\mathcal{A})).$$

Thus $(\mathcal{A},\pi) \to \Lambda^s(\mathcal{A})$ is, for each $s \geq 0$, a functor from sheaves with operators to sheaves. If π has order k, then denote by $(\mathcal{A},\pi) \to k^s\mathcal{A}$ the functor which to $(\mathcal{A},\pi)$ assigns the k^s-fold direct sum of $\mathcal{A}$ with itself.

(6.1) <u>Lemma</u>: <u>For each</u> $s \geq 0$ <u>the functors</u> $(\mathcal{A},\pi) \to \Lambda^s(\mathcal{A})$ <u>and</u> $(\mathcal{A},\pi) \to k^s\mathcal{A}$ <u>are naturally equivalent</u>.

Order once and for all the elements of π. For $s > 0$ use this to lexicographically order the generators of $\beta_s(\pi)$. Since $\beta_s(\pi)$ is a free $Z(\pi)$-module on the ordered s-tuples of elements of π there are just k^s generators. Thus each element in $\operatorname{Hom}_{Z(\pi)}(\beta_s(\pi),\Gamma(U,\mathcal{A}))$ is uniquely determined by an ordered k^s-tuple of sections in $\Gamma(U,\mathcal{A})$. This proves

$$\mathrm{Hom}_{Z(\pi)}(\beta_s(\pi),\Gamma(U,\mathcal{A}))\simeq k^s\,\Gamma(U,\mathcal{A})$$

and hence the lemma. The following lemma is now trivial.

(6.2) Lemma: For each $s \geq 0$,

1) $(\mathcal{A},\pi) \to \Lambda^s(\mathcal{A})$ preserves exactness

2) for $U \subset Y$

$$\mathrm{Hom}_{Z(\pi)}(\beta_s(\pi),\Gamma(U,\mathcal{A}))\simeq\Gamma(U,\Lambda^s(\mathcal{A}))$$

3) if $\mathcal{A}$ is flabby, so is $\Lambda^s(\mathcal{A})$

4) if $\mathcal{A}$ is acyclic, so is $\Lambda^s(\mathcal{A})$.

We are particularly concerned with the functor

$$(\mathcal{A},\pi)\to\Lambda(\mathcal{A}):\Lambda^0(\mathcal{A})\xrightarrow{d^0}\Lambda^1(\mathcal{A})\to\cdots\to\Lambda^s(\mathcal{A})\xrightarrow{d^s}\cdots$$

to the category of differential sheaves. We take $d^s:\Lambda^s(\mathcal{A})\to\Lambda^{s+1}(\mathcal{A})$ to be the dual homomorphism

$$\partial^*_{s+1}:\mathrm{Hom}_{Z(\pi)}(\beta_s(\pi),\Gamma(U,\mathcal{A}))$$
$$\longrightarrow\mathrm{Hom}_{Z(\pi)}(\beta_{s+1}(\pi),\Gamma(U,\mathcal{A})).$$

Associated to every differential sheaf there is a bigraded complex [5, p 128], which for $\Lambda(\mathcal{A})$ we denote by

$$C^{r,s}(\pi;\mathcal{A})=C^r(Y;\Lambda^s(\mathcal{A})).$$

In this situation

$$\delta':C^{r,s}(\pi;\mathcal{A})\longrightarrow C^{r+1,s}(\pi;\mathcal{A})$$

is induced from the canonical resolution

$$0 \to \mathcal{A}^s(\mathcal{A}) \to \mathcal{C}^o(Y;\mathcal{A}^s(\mathcal{A})) \to \dots \to \mathcal{C}^r(Y;\mathcal{A}^s(\mathcal{A})) \to \dots$$

while $\delta'' : C^{r,s}(\pi;\mathcal{A}) \to C^{r,s+1}(\pi;\mathcal{A})$ is induced from $\mathcal{A}^s(\mathcal{A}) \xrightarrow{(-1)^r d^s} \mathcal{A}^{s+1}(\mathcal{A})$ by application of the functorial property of the canonical resolution. Put $C^p(\pi;\mathcal{A}) = \sum_{r+s=p} C^{r,s}(\pi;\mathcal{A})$, $\delta = \delta' + \delta''$, then the resulting cohomology group is by definition $H^p(\pi;\mathcal{A})$. By we see that if $\mathcal{h}^s \to Y$ the sheaf of cohomology groups arising from

$$\dots \to \mathcal{A}^{s-1}(\mathcal{A}) \to \mathcal{A}^s(\mathcal{A}) \to \mathcal{A}^{s+1}(\mathcal{A}) \to \dots$$

then

(6.3) <u>Theorem</u>: <u>For any sheaf with operators</u> $(\mathcal{A},\pi) \to Y$ <u>there is a spectral sequence</u> $\{{}'E_t^{r,s} d_t\} \Rightarrow H^*(\pi;\mathcal{A})$ <u>with</u>

$${}'E_2^{r,s} \simeq H^r(Y;h^s).$$

This is simply the sheaf version of (5.1). We note that $\mathcal{h}^s \to Y$ is defined by the presheaf $U \to H^s(\pi,\Gamma(U,\mathcal{A}))$, and $\mathcal{h}^s_y = H^s(\pi;\mathcal{A}_y)$ for each $y \in Y$.

(6.4) <u>Theorem</u>: <u>To each short exact sequence of sheaves with operators</u>

$$0 \to (\mathcal{A}',\pi) \to (\mathcal{A},\pi) \to (\mathcal{A}'',\pi) \to 0$$

<u>there is naturally associated an exact coefficient sequence</u>

$$\ldots \to H^p(\pi;\mathcal{A}') \to H^p(\pi;\mathcal{A}) \to H^p(\pi;\mathcal{A}'') \to H^{p+1}(\pi;\mathcal{A}') \to \ldots$$

Since $\Lambda^s(\cdot)$ preserves exactness,

$$0 \to \Lambda^s(\mathcal{A}') \to \Lambda^s(\mathcal{A}) \to \Lambda^s(\mathcal{A}'') \to 0$$

is exact for each $s \geq 0$. But $C^r(Y;\cdot)$ also preserves exactness, hence

$$0 \to C^r(Y;\Lambda^s(\mathcal{A}')) \to C^r(Y;\Lambda^s(\mathcal{A})) \to C^r(Y;\Lambda^s(\mathcal{A}'')) \to 0$$

remains exact, whence the theorem.

Next we must consider a differential sheaf with operators

$$(\mathcal{L},\pi) : (\mathcal{L}^0,\pi) \to (\mathcal{L}^1,\pi) \to \ldots \to (\mathcal{L}^i,\pi) \to \ldots$$

The result of applying Λ to $(\mathcal{L},\pi)$ is a bigraded differential sheaf $\Lambda(\mathcal{L})$ with $\Lambda(\mathcal{L})^s = \sum_{i+j=s} \Lambda^j(\mathcal{L}^i)$. We take $d' : \Lambda^j(\mathcal{L}^i) \to \Lambda^j(\mathcal{L}^{i+1})$ to be induced by $(\mathcal{L}^i,\pi) \to (\mathcal{L}^{i+1},\pi)$ while $d'' : \Lambda^j(\mathcal{L}^i) \to \Lambda^{j+1}(\mathcal{L}^i)$ is given by

$$\begin{aligned}(-1)^i \partial^*_{j+1} : &\operatorname{Hom}_{Z(\pi)}(\beta_j(\pi), \Gamma(U,\mathcal{L}^i)) \\ \longrightarrow &\operatorname{Hom}_{Z(\pi)}(\beta_{j+1}(\pi), \Gamma(U,\mathcal{L}^i)).\end{aligned}$$

Put $d = d' + d''$ to obtain the bigraded differential sheaf

$$\Lambda(\mathcal{L}) : \Lambda(\mathcal{L})^0 \to \Lambda(\mathcal{L})^1 \to \ldots \to \Lambda(\mathcal{L})^s \to \ldots .$$

To compute the sheaf of cohomology groups $\mathcal{H}^*(\Lambda(\mathcal{L})) \to Y$ we note that at each $y \in Y$ there are two spectral sequences

$$\{{}'\mathcal{E}_t^{i,j}(y), d_t\} \Rightarrow H^*(\Lambda(\mathcal{L})y) \Leftarrow \{ {}''\mathcal{E}_t^{i,j}(y),\ d_t\}$$

where

$$'\mathcal{E}_2^{i,j}(y) = H^i(H^j(\pi; \mathcal{L}_y))$$

$$''\mathcal{E}_2^{j,i}(y) = H^j(\pi; H^i(\mathcal{L}_y)).$$

It is possible to think of two spectral sequences in which each $'\mathcal{E}_t^{i,j}$ (or $''\mathcal{E}_t^{i,j}$) is a sheaf over Y and d_t is a sheaf homomorphism. In other words $'\mathcal{E}_t^{i,j}$ is defined by the pre-sheaf

$$U \longrightarrow {}'\mathcal{E}_t^{i,j}(U)$$

where $\{{}'\mathcal{E}_t^{i,j}(U), d_t\}$ arises from the bigraded $\Gamma(U, \Lambda(\mathcal{L}))$. With this in mind

$$'\mathcal{E}_2^{i,j} \simeq \mathcal{H}^i(\mathcal{H}^j(\pi; \mathcal{L}))$$

$$''\mathcal{E}_2^{j,i} \simeq \mathcal{H}^j(\pi; \mathcal{H}^i(\mathcal{L})).$$

We proceed to introduce the bigraded complex K with

$$K^{r,s} = C^r(Y; \Lambda(\mathcal{L})^s)$$

where δ' is induced from the canonical resolution of $\Lambda(\mathcal{L})^s$ and $(-1)^r\delta''$ is induced by the differential of the sheaf $\Lambda(\mathcal{L})$ with the aid of the functorial property of the canonical resolution.

(6.5) Lemma: If

$$0 \to (\mathcal{A},\pi) \to (\mathcal{L}^0,\pi) \to (\mathcal{L}^1,\pi) \to \dots \to (\mathcal{L}^i,\pi)$$

is a resolution with operators then there is a natural isomorphism of $\mathcal{h}^s \to Y$ with the sheaf of cohomology groups $\mathcal{H}^s(\Lambda(\mathcal{L})) \to Y$.

We see that $\mathcal{H}^i(\mathcal{L}) = 0$, $i > 0$; thus from the "$\mathcal{E}$-spectral sequence

$$\mathcal{H}^s(\Lambda(\mathcal{L})) \simeq {}''\mathcal{E}_2^{s,0} \simeq \mathcal{H}^s(\pi;\mathcal{H}^0(\mathcal{L})).$$

But $(\mathcal{A},\pi) \to (\mathcal{L}^0,\pi)$ induces $(\mathcal{A},\pi) \simeq (\mathcal{H}^0(\mathcal{L}),\pi)$ so that $\mathcal{h}^s \simeq \mathcal{H}^s(\pi;\mathcal{H}^0(\mathcal{L})) \simeq \mathcal{H}^s(\Lambda(\mathcal{L}))$.

(6.6) Lemma: If

$$0 \to (\mathcal{A},\pi) \to (\mathcal{L}^0,\pi) \to \dots \to (\mathcal{L}^i,\pi) \to \dots$$

is a resolution with operators then the canonical embedding

$$j : C^{r,s}(\pi;\mathcal{A}) \to K^{r,s}$$

induces an isomorphism

$$H^p(\pi;\mathcal{A}) \simeq H^p(K).$$

The embedding is induced from

$$0 \to \Lambda^s(\mathcal{A}) \to \Lambda^s(\mathcal{L}^0) \to \dots \to \Lambda^s(\mathcal{L}^i) \to \dots$$

by way of

$$0 \longrightarrow C^r(Y;\Lambda^s(\mathcal{a})) \longrightarrow C^r(Y;\Lambda^s(\mathcal{L}^o)) \longrightarrow \dots .$$

Note that the image of $\Lambda^s(\mathcal{a})$ in $\Lambda^s(\mathcal{L}^o)$ is annihilated by $\Lambda^s(\mathcal{L}^o) \xrightarrow{d^o} \Lambda^s(\mathcal{L}^1)$, hence $\Lambda^s(\mathcal{a}) \to \Lambda(\mathcal{L})^s$ is the embedding of one differential sheaf into another, thus $C^r(Y;\Lambda^s(\mathcal{a})) \longrightarrow K^{r,s}$ is a cochain complex embedding. Incidentally, the sign choices are compatible. If, in the light of (6.5), we compare the $'E_2$-terms of the 'E-spectral sequences of $C^{r,s}(Y;\mathcal{a})$ and $K^{r,s}$ we see these $'E_2$-terms are in fact isomorphic, hence the lemma.

(6.7) Theorem: If

$$0 \longrightarrow (\mathcal{a},\pi) \longrightarrow (\mathcal{L}^o,\pi) \longrightarrow \dots \longrightarrow (\mathcal{L}^i,\pi) \longrightarrow \dots$$

is a resolution with operators for which each $\mathcal{L}^i \longrightarrow Y$ is an acyclic sheaf then

$$H^p(\pi;\mathcal{a}) \simeq H^p(\Gamma(Y,\Lambda(\mathcal{L})).$$

We know from (6.6) that $H^p(\pi;\mathcal{a}) \simeq H^p(K)$. Consider the "E-spectral sequence of K. In general, [5, p. 129],

$$"E_2^{p,q} \simeq H^p(H^q(Y;\Lambda(\mathcal{L}))).$$

We noted in (6.2) that Λ preserves the acyclicity of sheaves, hence $"E_2^{p,q} = 0$, $q > 0$ and $"E_2^{p,o} \simeq H^p(\Gamma(Y;\Lambda(\mathcal{L}))$ and thus (6.7) follows. According to (6.2), part 2,

$$\Gamma(Y,\Lambda^s(\mathcal{L}^r)) = \mathrm{Hom}_{Z(\pi)}(\beta_s(\pi),\Gamma(Y,\mathcal{L}^r)).$$

This means $\Gamma(Y,\mathcal{A}(\mathcal{L}))$ is really a bigraded cochain complex too with δ' given by $\Gamma(Y,\mathcal{L}^r) \to \Gamma(Y,\mathcal{L}^{r+1})$ and δ'' by $(-1)^r \partial^*_{s+1} : \mathrm{Hom}_{Z(\pi)}(\beta_s(\pi), \Gamma(Y,\mathcal{L}^r))$
$\longrightarrow \mathrm{Hom}_{Z(\pi)}(\beta_{s+1}(\pi), \Gamma(Y,\mathcal{L}^r))$. The sheaf analog of (5.2) is the "E-spectral sequence of this bigraded complex, for since $\mathcal{L}$ was an acyclic resolution of $\mathcal{Q}$,

$$(H^r(\Gamma(Y,\mathcal{L})),\pi) \simeq (H^r(Y;\mathcal{Q}),\pi).$$

(6.8) <u>Corollary</u>: <u>For any sheaf with operators</u> $(\mathcal{Q},\pi) \to Y$ <u>there is a spectral sequence</u> $\{{}''E_t^{s,r}, d_t\} \Rightarrow H^*(\pi;\mathcal{Q})$ <u>with</u>

$${}''E_2^{s,r} \simeq H^s(\pi;H^r(Y;\mathcal{Q})).$$

We have only to exhibit the canonical resolution of $(\mathcal{Q},\pi)$ with operators. Recall that

$$C^o(U,\mathcal{Q}) = \{f \mid f : U \to \mathcal{Q},\ pf = \text{identity}\}$$

where we do not require f to be continuous. A right $Z(\pi)$-module structure on $C^o(U;\mathcal{Q})$ is given by $(fh)(x) = f(x)\,h$. The presheaf $U \to (C^o(U,\mathcal{Q}),\pi)$ defines $(\mathcal{C}^o(Y;\mathcal{Q}),\pi)$. Since $(\mathcal{Q},\pi) \subset (\mathcal{C}^o(Y,\mathcal{Q}),\pi)$ it follows that $\mathcal{F}^1(Y;\mathcal{Q}) = \mathcal{C}^o(Y;\mathcal{Q})/\mathcal{Q}$ is also a sheaf with operators. By repeating the construction we arrive at the canonical resolution with operators

$$0 \to (\mathcal{Q},\pi) \to (\mathcal{C}^o(Y;\mathcal{Q}),\pi) \to \dots \to \mathcal{C}^r(Y;\mathcal{Q}),\pi) \to \dots$$

which is certainly an acyclic resolution.

Suppose that π acts trivially on $\mathcal{A}$, then π acts trivially on each $\mathcal{C}^r(Y;\mathcal{A})$ and

$$C^r(Y;\mathcal{A}) = \Gamma(Y;\mathcal{C}^r(Y;\mathcal{A}))$$

is a trivial $Z(\pi)$-module so in this case

$$C^{r,s}(\pi;\mathcal{A}) = \mathrm{Hom}_{Z(\pi)}(\beta_s(\pi), C^r(Y;\mathcal{A})) \simeq \mathrm{Hom}_Z(\beta_s(\pi) \otimes_{Z(\pi)} Z, C^r(Y;\mathcal{A}))$$

Thus if π acts trivially on $\mathcal{A}$ we can compute $H^p(\pi;\mathcal{A})$ by means of the Künneth formula. We elect to present the answer in the following form

(6.9) Corollary: If $\mathcal{A} \to Y$ is a sheaf with trivial operators there is a (split) short exact sequence

$$0 \to \sum_{r+s=p} H^r(Y;\mathcal{A}) \otimes H^s(\pi;Z) \to H^p(\pi;\mathcal{A}) \to$$

$$\to \sum_{r+s=p+1} \mathrm{Tor}\,(H^r(Y;\mathcal{A}), H^s(\pi;Z)) \to 0.$$

This is the one case in which we can get a complete answer for $H^*(\pi;\mathcal{A})$.

7. Nerves of coverings

We fix a locally finite open covering $\mathcal{V}$ of Y made up of π-invariant open sets and K denotes the nerve of $\mathcal{V}$. To $\sigma \in \mathcal{K}$ assign the π-invariant, non-empty, open set Sup $(\sigma) \subset Y$, the intersection of the vertices making up σ. If $(\mathcal{A}, \pi) \to Y$ is a sheaf with operators, then a covariant stack with operators $\mathcal{A}(\mathcal{V}) \to K$ is given by assigning to each $\sigma \in \mathcal{K}$ the right $Z(\pi)$-module of sections $\Gamma(\text{Sup}\ (\sigma); \mathcal{A})$.

We want to define a natural homomorphism of $H^*(\pi; \mathcal{A}(\mathcal{V})) \to H^*(\pi; \mathcal{A})$ in order to relate the definitions of sections 5 and 6. Let $(\mathcal{L}, \pi)$ be the canonical resolution with operators

$$0 \to (\mathcal{A}, \pi) \to (\mathcal{L}^0, \pi) \to \ldots . \to (\mathcal{L}^i, \pi) \to \ldots$$

and introduce again the bigraded differential sheaf $\Lambda(\mathcal{L}) \to Y$. We use this to define the bigraded complex

$$K^{r,s}(\mathcal{V}) = C^r(K; \Lambda(\mathcal{L})^s),$$

which can also be written as

$$K^{r,s}(\mathcal{V}) = \sum_{i+j=s} \text{Hom}_{Z(\pi)}(\beta_j(\pi), C^r(K; \mathcal{L}^i)).$$

Just as in section 6, there is a canonical embedding $C^{r,s}(\pi; \mathcal{A}(\mathcal{V})) \subset K^{r,s}(\mathcal{V})$ given by

$$\text{Hom}_{Z(\pi)}(\beta_s(\pi), C^r(K; \mathcal{A})) \to \text{Hom}_{Z(\pi)}(\beta_s(\pi), C^r(K; \mathcal{L}^0))$$

or equivalently

$$C^r(K;\Lambda^s(\mathcal{A})) \to C^r(K;\Lambda^s(\mathcal{L}^o)).$$

In any case, this induces a homomorphism

$$H^*(\pi;\mathcal{A}(\mathcal{V})) \to H^*(K(\mathcal{V})).$$

(7.1) Lemma: For $p \geq 0$

$$H^p(K(\mathcal{V})) \simeq H^p(\pi;\mathcal{A}).$$

We need only consider the "E-spectral sequence for $K(\mathcal{V})$. From general considerations, [9],

$$"E_2^{s,r} \simeq H^s(H^r(K;\Lambda(\mathcal{L}))).$$

Now each $\mathcal{L}^i$ is flabby, thus so is $\Lambda^j(\mathcal{L}^i)$, so according to [9],

$$H^r(K;\Lambda(\mathcal{L})) = 0$$

for $r > 0$ and $H^o(K;\Lambda(\mathcal{L})) = \Gamma(Y,\Lambda(\mathcal{L}))$. Therefore, $H^p(K(\mathcal{V})) \simeq H^p(\Gamma(Y;\Lambda(\mathcal{L})) \simeq H^p(\pi;\mathcal{A})$ by (6.7). Thus we have our natural homomorphism

$$H^*(\pi;\mathcal{A}(\mathcal{V})) \to H^*(\pi;\mathcal{A})$$

induced by $C^{r,s}(\pi;\mathcal{A}(\mathcal{V})) \to K^{r,s}(\mathcal{V})$.

(7.2) Theorem: If for each simplex σ in K

$$H^r(\mathrm{Sup}\,(\sigma);\mathcal{A}) = 0,\ r > 0$$

then $H^*(\pi;\mathcal{A}(\mathcal{V})) \simeq H^*(\pi;\mathcal{A})$.

The 'E-spectral sequence for $H^*(K(\mathcal{V}))$ begins with

$$'E_2^{r,s} \simeq H^r(K;\mathcal{H}^s(\Lambda(\mathcal{L})))$$

where $\mathcal{H}^s(\Lambda(\mathcal{L})) \to K$ is the covariant stack $\sigma \longrightarrow H^s(\Gamma(\mathrm{Sup}\ (\sigma),\Lambda(\mathcal{L})))$, [9]. We must show that

$$H^s(\pi;\Gamma(\mathrm{Sup}\ (\sigma),\mathcal{A})) \simeq H^s(\Gamma(\mathrm{Sup}\ (\sigma),\Lambda(\mathcal{L})).$$

Since $\mathcal{L}$ is the canonical resolution of $\mathcal{A}$ our hypothesis implies that

$$0 \to \Gamma(\mathrm{Sup}\ (\sigma),\mathcal{A}) \to \Gamma(\mathrm{Sup}\ (\sigma),\mathcal{L}^o) \to \ldots \to \Gamma(\mathrm{Sup}\ (\sigma),\mathcal{L}^i) \to \ldots$$

is exact. Using the fact that $\beta_j(\pi)$ is a free $Z(\pi)$-module and that

$$\begin{gathered}\mathrm{Hom}_{Z(\pi)}(\beta_j(\pi),\Gamma(\mathrm{Sup}\ (\sigma),\mathcal{L}^i))\\ \Gamma(\mathrm{Sup}\ (\sigma),\Lambda^j(\mathcal{L}^i))\end{gathered}$$

we conclude that

$$0 \to \Gamma(\mathrm{Sup}\ (\sigma),\Lambda^j(\mathcal{A})) \to \Gamma(\mathrm{Sup}\ (\sigma),\Lambda^j(\mathcal{L}^o)) \to \ldots.$$
$$\longrightarrow \Gamma(\mathrm{Sup}\ (\sigma),\Lambda^j(\mathcal{L}^i)) \longrightarrow \ldots.$$

is still exact. The required isomorphism follows immediately from the "E-spectral sequence for $H^*(\Gamma(\mathrm{Sup}\ (\sigma),\Lambda(\mathcal{L}))$ since

$$'H^r(\mathrm{Sup}\ (\sigma), \mathcal{A}^s(\mathcal{L})) = 0,\ r > 0$$

$$'H^o(\mathrm{Sup}\ (\sigma), \mathcal{A}^s(\mathcal{L})) \simeq \Gamma(\mathrm{Sup}\ (\sigma), \mathcal{Q})$$

so $H^s(\Gamma(\mathrm{Sup}\ (\sigma), \mathcal{A}(\mathcal{L})) \simeq H^s(\pi; \Gamma(\mathrm{Sup}\ (\sigma); \mathcal{Q}))$

With this last we simply compare the 'E-spectral sequences for $C^*(\pi; \mathcal{Q}(\mathcal{V}))$ and $K(\mathcal{V})$ and see that the $'E_2$-terms are isomorphic.

For a sheaf with operators $(\mathcal{Q}, \pi) \rightarrow Y$ we can of course give a Čech definition of $H^p(\pi; \mathcal{Q})$. The reader is requested to do this using the definitions of section 3.

8. The sheaves $p : (\mathcal{A},\pi) \to (\pi, X)$

So far we have only considered sheaves with operators over a base space on which π acts trivially. More generally a sheaf with operators $p : (\mathcal{A},\pi) \to (\pi, X)$ consists of a sheaf together with an assignment to each $h \in \pi$ of a sheaf cohomorphism

$$\begin{array}{ccc} \mathcal{A} & \xleftarrow{\;h^{\#}\;} & \mathcal{A} \\ {\scriptstyle p}\downarrow & & \downarrow{\scriptstyle p} \\ X & \xrightarrow{\;h\;} & X \end{array}$$

such that

1) $e^{\#} = \text{identity}$
2) $(h\hat{h})^{\#} = \hat{h}^{\#} \circ h^{\#}$

In terms of a pre-sheaf $U \to \mathcal{A}_U$ there is associated a family of isomorphisms

$$h^{\#}_U : \mathcal{A}_{hU} \to \mathcal{A}_U$$

commuting with the restrictions and for which $e^{\#}_U = \text{identity}$ and $(h\hat{h})^{\#}_U = \hat{h}^{\#}_U \circ h^{\#}_U$.

For $U \subset X$ let $K \subset \pi$ be the subgroup of elements with $hU = U$. There is a right $Z(K)$-module structure on $\Gamma(U,\mathcal{A})$ given by

$$(sk)(x) = k^{\#}s(kx)$$

(8.1) <u>**Theorem:**</u> <u>Suppose $U \subset X$ is an open subset with the</u>

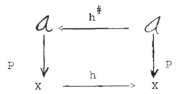

<u>property</u> $hU \cap U \neq \emptyset$ <u>if and only if</u> $hU = U$. <u>If</u> $V = \bigcup_{h \in \pi} hU$ <u>then there is a natural</u> $Z(\pi)$-<u>module isomorphism</u>

$$\mathrm{Hom}_{Z(K)}(Z(\pi), \Gamma(U,\mathcal{A})) \cong \Gamma(V;\mathcal{A}).$$

Here K is the subgroup of π with $hU = U$. An element of $\mathrm{Hom}_{Z(K)}(Z(\pi), \Gamma(U,\mathcal{A}))$ is a function which to each $h \in \pi$ assigns a section $s(h,x) \in \Gamma(U,\mathcal{A})$ so that $s(hk,x) = k^{\#}s(h,kx)$. A section $\sigma \in \Gamma(V,\mathcal{A})$ is defined by assigning to hx the unique element $\sigma(hx) \in \mathcal{A}_{hx}$ for which $h^{\#}\sigma(hx) = s(h,x) \in \mathcal{A}_x$. Suppose $hx = \hat{h}y$, $x,y \in U$, then $h^{-1}\hat{h} = k \in K$, $hky = hx$ and $ky = x$. Then $(hk)^{\#}\sigma(hky) = s(hk,y) = k^{\#}s(h,ky) = k^{\#}s(h,x) = k^{\#}h^{\#}\sigma(hx) = (hk)^{\#}\sigma(hx)$, hence $\sigma(\hat{h}y) = \sigma(hx)$ and σ is well defined. The $Z(\pi)$-module structure in terms of the $s(h,x)$ is $s(h,x)\hat{h} = s(\hat{h}h,x) = (\hat{h}h)^{\#}\sigma(\hat{h}hx) = h^{\#}\hat{h}^{\#}\sigma(\hat{h}hx) = h^{\#}(\sigma\hat{h}(hx))$ so we have a $Z(\pi)$-module homomorphism.

Conversely, given $\sigma \in \Gamma(V,\mathcal{A})$ define $s(h,x)$ to be $h^{\#}\sigma(hx)$ for all $x \in U$, $h \in \pi$. Then

$$s(hk,x) = (hk)^{\#}\sigma(hkx) = k^{\#}(h^{\#}\sigma(hkx)) = k^{\#}s(h,kx)$$

as required. This establishes the theorem.

Now let $\nu: X \to X/\pi$ be the quotient map. We define a direct image sheaf with operators $(\mathcal{A}^*, \pi) \to X/\pi$ to be induced by the pre-sheaf $V \to (\Gamma(\nu^{-1}(V), \mathcal{A}), Z(\pi))$ for every open set $V \subset X/\pi$.

(8.2) <u>Corollary</u>: <u>If</u> $\nu(x) = y$ <u>there is a</u> $Z(\pi)$-<u>module</u>

<u>isomorphism</u>

$$\mathrm{Hom}_{Z(\pi_x)}(Z(\pi),\, \mathcal{A}_x) \simeq \mathcal{A}^*_y \, .$$

We can choose arbitrarily small open sets U_x which are π_x-invariant and for which $hU_x \cap U_x \neq \emptyset \Longleftrightarrow h \in \pi_x$. Let $V_y \subset X/\pi$ be the open set determined by $\nu^{-1}(V_y) = \bigcup_{h \in \pi} hU_x$.

Then there is a $Z(\pi)$-module isomorphism

$$\mathrm{Hom}_{Z(\pi_x)}(Z(\pi),\, \Gamma(U_x, \mathcal{A})) \simeq \Gamma(V_y, \mathcal{A}^*)$$

and if $U'_x \subset U_x$ the diagram

$$\begin{array}{ccc} \mathrm{Hom}_{Z(\pi_x)}(Z(\pi),\, \Gamma(U_x, \mathcal{A})) & \simeq & \Gamma(V_y, \mathcal{A}) \\ \downarrow & & \downarrow \\ \mathrm{Hom}_{Z(\pi_x)}(Z(\pi),\, \Gamma(U'_x, \mathcal{A})) & \simeq & \Gamma(V'_y, \mathcal{A}) \end{array}$$

commutes. The corollary will follow if

$$\mathrm{Hom}_{Z(\pi_x)}(Z(\pi),\, \mathrm{dir\,lim}\, \Gamma(U_x, \mathcal{A})) \simeq \mathrm{dir\,lim}\, \mathrm{Hom}_{Z(\pi_x)}(Z(\pi),\, \Gamma(U_x, \mathcal{A}))$$

This is true since $Z(\pi)$ is a free right $Z(\pi_x)$-module. An appropriate basis is given by choosing a representative from each coset of π_x in π.

(8.3) <u>Corollary</u>: <u>If</u>

$$\begin{array}{ccccccccc} 0 & \longrightarrow & (\mathcal{A}_1, \pi) & \longrightarrow & (\mathcal{A}, \pi) & \longrightarrow & (\mathcal{A}_2, \pi) & \longrightarrow & 0 \\ & & & \searrow^{p'} & \downarrow p & \swarrow^{p''} & & & \\ & & & & (\pi, X) & & & & \end{array}$$

<u>is an exact sequence of sheaves with operators then</u>

$$0 \to (\mathcal{A}_1^*, \pi) \to (\mathcal{A}^*, \pi) \to (\mathcal{A}_2^*, \pi) \to 0$$

<u>is also an exact sequence of sheaves with operators on</u> X/π.

We are faced with two possibilities for defining $H^p(\pi;\mathcal{A})$. We might of course say that $H^p(\pi;\mathcal{A}) = H^p(\pi;\mathcal{A}^*)$, so that we get the definition from section 6. On the other hand we can surely take the canonical resolution with operators
$0 \to (\mathcal{A},\pi) \to (\mathcal{C}^o(X;\mathcal{A}),\pi) \to \dots \to (\mathcal{C}^r(X;\mathcal{A}),\pi) \to \dots,$
just as in section 6. Now by definition

$$C^r(X;\mathcal{A}) = \Gamma(X, \mathcal{C}^r(X;\mathcal{A}))$$

so that $C^r(X;\mathcal{A})$ is a right $Z(\pi)$-module. We put

$$C^{r,s}(\pi;\mathcal{A}) = \mathrm{Hom}_{Z(\pi)}(\beta_s(\pi), C^r(X;\mathcal{A}))$$

and take $H^p(\pi;\mathcal{A})$ to be the cohomology of the bigraded complex

$$C^p(\pi;\mathcal{A}) = \sum_{r+s=p} C^{r,s}(\pi;\mathcal{A}).$$

The two definitions are equivalent. In fact if

$$0 \to (\mathcal{A},\pi) \to (\mathcal{L}^o,\pi) \to \dots \to (\mathcal{L}^r,\pi) \to \dots$$

is a resolution by acyclic sheaves with operators, then $(\mathcal{L}^*,\pi)$ is still, by (8.3), a resolution of $(\mathcal{A}^*,\pi)$. In addition, according to [5, p. 54, II.1], $\mathcal{L}^*$ is an acyclic resolution. Now $\Gamma(X/\pi,\mathcal{L}^*) \simeq \Gamma(X,\mathcal{L})$ as $Z(\pi)$-modules. Thus
$H^*(\mathrm{Hom}_{Z(\pi)}(\beta_*(\pi), \Gamma(X/\pi,\mathcal{L}^*)) \simeq H^*(\mathrm{Hom}_{Z(\pi)}(\beta_*(\pi), \Gamma(X,\mathcal{L}))).$

By (6.7), however

$$H^*(\pi;\mathcal{A}^*) \simeq H^*(\mathrm{Hom}_{Z(\pi)}(\beta_*(\pi), \Gamma(X/\pi, \mathcal{L}^*))).$$

Roughly speaking, if we want to consider the "E-spectral sequence then we use the direct definition of $H^*(\pi;\mathcal{A})$ for then it is apparent that

$$''E_2^{s,r} \simeq H^s(\pi;H^r(X;\mathcal{A})).$$

On the other hand, for the 'E-spectral sequence we use $H^*(\pi;\mathcal{A}^*)$. In this case

$$'E_2^{r,s} \simeq H^r(X/\pi;\mathcal{h}^s)$$

were $\mathcal{h}^s \longrightarrow X/\pi$ is the sheaf defined by the pre-sheaf $V \longrightarrow H^s(\pi;\Gamma(\nu^{-1}(V),\mathcal{A}))$.

(8.4) <u>Corollary</u>: <u>For</u> $y = \nu(x)$ <u>there is a canonical isomorphism</u>

$$\mathcal{h}_y^s \simeq H^s(\pi_x, \mathcal{A}_x) \simeq H^s(\pi;\mathcal{A}_y^*).$$

We can now state the fundamental Borel-Cartan result.

(8.5) <u>Theorem</u>: <u>If</u> $(\mathcal{A},\pi) \longrightarrow (\pi,X)$ <u>is a sheaf with operators such that at each</u> $x \in X$, $H^p(\pi_x;\mathcal{A}_x) = 0$, $p > 0$, <u>then there is a cohomology spectral sequence</u> $\{E_2^{s,r}d_t\} \Longrightarrow H^*(X/\pi;\mathcal{h}^o)$ <u>with</u>

$$E_2^{s,r} \simeq H^s(\pi;H^r(X;\mathcal{A})).$$

Clearly the "E-spectral sequence for $H^*(\pi;\mathcal{A})$ is what we need. But the 'E-spectral sequence shows $H^*(X/\pi;\mathcal{h}^0) \simeq H^*(\pi;\mathcal{A})$.

9. Topological examples

Fix a pair (π, G) consisting of a left action of π on an abelian topological group G. For each $U \subset X$ open let $\mathcal{G}(U)$ be the abelian group of all maps $f : U \to G$. Define $h_U^{\#} : \mathcal{G}(hU) \to \mathcal{G}(U)$ by $h_U^{\#}(f)(x) = h^{-1}f(hx)$. Let $(\mathcal{G}, \pi) \to (\pi, X)$ denote the resulting sheaf with operators over (π, X). By appealing to the Čech definition we can regard $H^1(\pi; \mathcal{G})$ as the abelian group of equivalence classes of right principal (π, G) spaces over (π, X).

For example, take a unimodular representation of π on C^k, then $Z^k \subset C^k$ is invariant and from

$$0 \to Z^k \to C^k \to G \to 0$$

we induce an action of π on $G = (C^*)^k$ as a group of automorphisms. We define sheaves with operators $(\mathcal{Z}^k, \pi) \to (\pi, X)$ and $(\mathcal{C}^k, \pi) \to (\pi, X)$ by analogy with $(\mathcal{G}, \pi) \to (\pi, X)$. In fact, $\mathcal{Z}^k \to X$ is just the k-fold direct sum of the constant sheaf $X \times Z \to X$ with itself. We shall assume (a local condition on X) that

$$0 \to (\mathcal{Z}^k, \pi) \to (\mathcal{C}^k, \pi) \to (\mathcal{G}, \pi) \to 0$$

is exact. We can then, with the aid of (6.4), show that

$$H^1(\pi; \mathcal{G}) \simeq H^2(\pi; \mathcal{Z}^k).$$

We must show, in other words, that $H^p(\pi; \mathcal{C}^k) = 0$, $p > 0$. In the "E-spectral sequence for $H^*(\pi; \mathcal{C}^k)$ we have $"E_2^{s,r} = 0$, $r > 0$ because $\mathcal{C}^k$ is a fine sheaf and $H^r(X; \mathcal{C}^k) = 0$, $r > 0$. Furthermore, $"E_2^{s,0} \simeq H^s(\pi; H^0(X; \mathcal{C}^k)) = 0$ for $s > 0$ since π is finite and $H^0(X; \mathcal{C}^k)$ is a complex vector space. The isomorphism $H^1(\pi; \mathcal{G}) \to H^2(\pi; \mathcal{Z}^k)$ may be interpreted as assigning to each right principal (π, G)-space its Chern class.

It is helpful to recognize that $H^p(X; \mathcal{Z}^k)$ is really a $Z(\pi) - Z(\pi)$ bimodule. The right action of π is induced naturally from the action of π on X, while the left action of π is due to the unimodular representation of π on Z^k. We have reduced the bimodule structure to a right $Z(\pi)$-module in the customary fashion. Suppose, for example, that X is connected and $H^1(X;Z) = H^2(X;Z) = 0$. Then from the "E-spectral sequence we see $H^2(\pi; \mathcal{Z}^k) \simeq H^2(\pi; H^0(X; \mathcal{Z}^k))$. Now $H^0(X; \mathcal{Z}^k) \simeq Z^k$ and the action of π on X induces a trivial action on Z^k, thus the entire $Z(\pi)$-module structure is due to the unimodular representation. Therefore we can write

$$H^1(\pi; \mathcal{G}) \simeq H^2(\pi; Z^k) \simeq H^1(\pi; G)$$

in this case.

Another possibility is to study the complex line bundles with operators over (π, X). In this case, $G = C^*$ and the action of π on C^* is trivial. We write $H^1(\pi; \mathcal{C}^*) \simeq H^2(\pi; \mathcal{Z})$ in this case. It is suitable to refer to line bundles with operators here for if (π, B, C^*) is a right principal (π, C^*) space then we form the complex line bundle

$$\xi = (B \times C)/C^{*} \longrightarrow X$$

and since π acts trivially on C^{*} the action $(\pi, \xi) \longrightarrow (\pi, X)$ is given by $h((b, z)) = ((hb, z))$.

Still another example is furnished by the Atiyah real line bundles, [1] . Take $\pi = Z_2$ and $G = U(1)$ and let Z_2 act on $U(1)$ by the conjugation automorphism, then $H^1(Z_2; \mathcal{U}(1))$ is the group of Atiyah real line bundles over (Z_2, X); that is, given a $(Z_2, B, U(1))$ form

$$\xi = (B \times C)/U(1) \longrightarrow X$$

as usual, then on ξ define a conjugation involution by

$$t((b, z)) = ((T(b), \bar{z})).$$

Chern classes are seen to exist as follows. There is

$$0 \longrightarrow Z \longrightarrow R \longrightarrow U(1) \longrightarrow 0$$

where Z_2 acts on R (and hence Z) by $r \longrightarrow -r$. This makes a sheaf with operators $(\mathcal{Z}, Z_2) \longrightarrow (Z_2, X)$ and $H^1(Z_2; \mathcal{U}(1)) \simeq H^2(Z_2, \mathcal{Z})$. If we remember to think of $H^{*}(X;Z)$ as a $Z(Z_2) - Z(Z_2)$ bimodule it is not hard to compute the "$E_2^{s,r}$-terms for $H^{*}(Z_2; \mathcal{Z})$ in this case. Problem: Determine the '$E_2^{r,s} \simeq H^r(X/Z_2; \mathcal{h}^s)$ in this case.

This will serve to indicate how the cohomology of a group with coefficients in a sheaf with operators relates to the study of right principal (π, G)-spaces for G an abelian topological group.

10. Holomorphic line bundles with operators

Let V^n be a closed connected Kahler manifold. We shall briefly recall the work of Kodaira and Spencer on the Picard group $\mathcal{P}(V^n)$, [15]. Denote by $\mathcal{O}(V^n)$, $\mathcal{O}^*(V^n)$ respectively the sheaves of germs of holomorphic functions and of non-vanishing holomorphic functions. There is a sheaf homomorphism $e : \mathcal{O}(V^n) \to \mathcal{O}^*(V^n) \to 0$ given again by $e(f) = \exp(2\pi i f)$. If $\mathcal{Z}(V^n)$ is the constant sheaf, then there is an exact sequence

$$0 \to \mathcal{Z}(V^n) \to \mathcal{O}(V^n) \xrightarrow{e} \mathcal{O}^*(V^n) \to 0$$

which produces the long cohomology sequence

$$\ldots \to H^p(V^n;Z) \to H^p(V^n;\mathcal{O}(V^n)) \to H^p(V^n;\mathcal{O}^*(V^n)) \to H^{p+1}(V^n;Z) \to \ldots.$$

By Dolbeault's Theorem, [5,p.126],
$H^p(V^n;\mathcal{O}(V^n)) \simeq h^{o,p}(V^n)$. Using the fact that V^n is closed and connected $H^o(V^n;\mathcal{O}^*(V^n)) = \Gamma(V^n;\mathcal{O}^*(V^n)) \simeq C^*$. Thus $e^* : H^o(V;\mathcal{O}(V^n)) \to H^o(V^n;\mathcal{O}^*(V^n))$ is onto. Kodaira and Spencer then concern themselves with

$$0 \to H^1(V^n;Z) \to H^1(V^n;\mathcal{O}(V^n)) \xrightarrow{e^*} H^1(V^n;\mathcal{O}^*(V^n)) \xrightarrow{c} H^2(V^n;Z).$$

Of course they regard $H^1(V^n;\mathcal{O}^*(V^n))$ as the abelian group of holomorphic equivalence classes of the holomorphic line bundles on V^n and $c : H^1(V^n;\mathcal{O}^*(V^n)) \to H^2(V^n;Z)$ assigns to each line bundle its Chern class. The Picard group

$\mathcal{P}(V^n) \subset H^1(V^n; \mathcal{O}^*(V^n))$ is defined to be the kernel of c. Thus $\mathcal{P}(V^n)$ is the group of holomorphic line bundles which are topologically trivial. By exactness, $\mathcal{P}(V^n)$ is naturally isomorphic to the quotient $H^1(V^n; \mathcal{O}(V^n))/H^1(V^n; Z)$. The Kähler assumption proves that $\mathcal{P}(V^n)$ is compact. That is, $H^1(V^n; C) \simeq h^{o,1}(V^n) \oplus h^{1,o}(V^n)$, and $h^{o,1}(V^n) \simeq h^{1,o}(V^n)$, hence

$$\operatorname{rank} H^1(V^n; Z) = 2 \dim_c H^1(V^n; \mathcal{O}(V^n)) = 2 \dim_c h^{o,1}(V^n).$$

The quotient of the vector group $H^1(V^n; \mathcal{O}(V^n))$ by the discrete subgroup $H^1(V^n; Z)$ is seen to be a compact, connected complex analytic abelian group; namely, $\mathcal{P}(V^n)$.

Suppose now that we add the action (π, V^n) of π as a group of holomorphic isometries of the Kahler metric. Clearly, $\mathcal{Z}(V^n)$, $\mathcal{O}(V^n)$ and $\mathcal{O}^*(V^n)$ all receive operator structures and the sequence

$$0 \to (\mathcal{Z}(V^n), \pi) \to (\mathcal{O}(V^n), \pi) \xrightarrow{e} \mathcal{O}^*(V^n, \pi) \to 0$$

is still exact, yielding now

$$\ldots \to H^p(\pi; \mathcal{Z}(V^n)) \to H^p(\pi; \mathcal{O}(V^n)) \xrightarrow{e^*} H^p(\pi; \mathcal{O}^*(V^n))$$

$$\to H^{p+1}(\pi; \mathcal{Z}(V^n)) \to \ldots \quad .$$

We can immediately determine $H^p(\pi; \mathcal{O}(V^n))$ since "$E_2^{s,r} \simeq H^s(\pi; H^r(V^n; \mathcal{O}(V^n)))$, but π is finite, while $H^r(V^n; \mathcal{O}(V^n))$ is a complex vector space, hence

$$H^p(\pi; \mathcal{O}(V^n)) \simeq {}''E_2^{o,p} \simeq H^p(V^n; \mathcal{O}(V^n))^{\pi} .$$

Similarly, we can show that $H^1(\pi;\mathcal{O}(V^n)) \simeq H^1(V^n;Z)^{\pi}$ because "$E_2^{1,0} \simeq H^1(\pi;H^0(V^n;Z)) \simeq \mathrm{Hom}\,(\pi,Z)$ is trivial. We argue that $H^0(\pi;\mathcal{O}^*(V^n)) \simeq C^*$, thus $e^* : H^0(\pi;\mathcal{O}(V^n)) \to H^0(\pi;\mathcal{O}^*(V^n))$ is still onto.

This results in a commutative diagram

$$\begin{array}{ccccccc}
H^1(V^n;Z) & \to & H^1(V^n;\mathcal{O}(V^n)) & \to & H^1(V^n;\mathcal{O}^*(V^n)) & \xrightarrow{c} & H^2(V^n;Z) \\
\uparrow & & \uparrow & & & & \\
H^1(V^n;Z)^{\pi} & \to & H^1(V^n;\mathcal{O}(V^n))^{\pi} & \to & H^1(\pi;\mathcal{O}^*(V^n)) & \xrightarrow{c} & H^2(\pi;\mathcal{Z}(V^n))
\end{array}$$

By analogy, let $\mathcal{P}(\pi,V^n) \subset H^1(\pi;\mathcal{O}^*(V^n))$ be the kernel of $c : H^1(\pi;\mathcal{O}^*(V^n)) \to H^2(\pi;\mathcal{Z}(V^n))$. We regard $H^1(\pi;\mathcal{O}^*(V^n))$ as the abelian group of holomorphic line bundles with operators and $\mathcal{P}(\pi,V^n)$ is the subgroup of those, which, as bundles with operators, are topologically equivalent to the trivial line bundle with operators.

(10.1) <u>Theorem</u>: <u>The Picard group $\mathcal{P}(\pi,V^n)$ is a closed connected complex analytic subgroup of $\mathcal{P}(V^n)$.</u>

The reader may show that $\mathrm{rank}\, H^1(V^n;Z)^{\pi} = 2\dim_c H^1(V^n;\mathcal{O}(V^n))^{\pi}$ which proves $\mathcal{P}(\pi,V^n)$ is compact. Note that if $v \in H^1(V^n;Z)$ has $i^*(v) \in H^1(V^n;\mathcal{O}(V^n))^{\pi}$ then $i^*(v-vh) = 0$ for all $h \in \pi$, but i^* is a monomorphism, so $v \in H^1(V^n;Z)^{\pi}$ and hence $\mathcal{P}(\pi,V^n) \subset \mathcal{P}(V^n)$ as asserted.

Of course π acts from the right on $\mathcal{P}(V^n)$ as a finite group of automorphisms. Let $\mathcal{P}(V^n)^{\pi} \subset \mathcal{P}(V^n)$ denote the subgroup of elements fixed under the action of every $h \in \pi$.

(10.2) <u>Corollary</u>: <u>The quotient group</u> $\mathcal{P}(V^n)^{\pi}/\mathcal{P}(\pi,V^n)$ <u>is isomorphic to</u> $H^1(\pi;H^1(V^n;Z))$.

There is an exact sequence of $Z(\pi)$-modules

$$0 \longrightarrow H^1(V^n;Z) \longrightarrow H^1(V^n;\mathcal{O}(V^n)) \longrightarrow \mathcal{P}(V^n) \longrightarrow 0.$$

Since π is finite and $H^1(V^n;\mathcal{O}(V^n))$ is a complex vector space

$$H^p(\pi;\mathcal{P}(V^n)) \simeq H^{p+1}(\pi;H^1(V^n;Z))$$

for $p > 0$, and the sequences

$$0 \longrightarrow H^1(V^n;Z)^{\pi} \longrightarrow H^1(V^n;\mathcal{O}(V^n))^{\pi}$$
$$\longrightarrow \mathcal{P}(V^n)^{\pi} \longrightarrow H^1(\pi;H^1(V^n;Z)) \longrightarrow 0$$

is exact. Since $\mathcal{P}(\pi,V^n) = H^1(V^n;\mathcal{O}(V^n))^{\pi}/H^1(V^n;Z)^{\pi}$, the corollary follows. While $H^1(V^n;Z)$ has no torsion, we cannot conclude in general that $H^1(\pi;H^1(V^n;Z)) = 0$ because $H^1(V^n;Z)$ may have a non-trivial $Z(\pi)$-module structure.

Define the sheaf $\mathcal{O}^*(V^n/\pi) \longrightarrow V^n/\pi$ by the pre-sheaf which assigns to $V \subset V^n/\pi$ the abelian group of holomorphic functions $f : \nu^{-1}(V) \longrightarrow C^*$ for which $f(hx) \equiv f(x)$, all $h \in \pi$. Thus $\mathcal{O}^*(V^n/\pi) \longrightarrow \hat{V}/\pi$ is just the sheaf $\mathcal{h}^0 \longrightarrow \hat{V}/\pi$ which appears in the 'E-spectral sequence for $H^*(\pi;\mathcal{O}^*(V^n))$. Although V^n/π is not in general a manifold, it is an analytic space and $\mathcal{O}^*(V^n/\pi)$ is the sheaf of germs of non-vanishing holomorphic functions on V^n/π. It is correct to regard $H^1(V^n/\pi;\mathcal{O}^*(V^n/\pi))$ as the holomorphic line bundles over the quotient space. Suppose that we have defined sheaves $\mathcal{O}(V^n/\pi)$ and $\mathcal{Z}(V^n/\pi)$ by analogy, then $\mathcal{Z}(V^n/\pi)$ is still the constant sheaf $(V^n/\pi) \times Z \to V^n/\pi$. If we expect to use the 'E-spectral sequences we should have an interpretation of $\mathfrak{h}^p \to \tilde{\mathcal{V}}/\pi$.

(10.3) <u>Lemma</u>: <u>If $\nu(x) = y$, then there is a natural isomorphism</u>

$$\mathfrak{h}^p_y \simeq H^p(\pi_x;C)$$

for all $p > 0$.

We can choose an arbitrarily small open π_x-invariant set U_x such that

1) U_x is connected and $H^1(U_x;Z) = 0$

2) $hU_x \cap U_x \neq \emptyset$ if and only if $h \in \pi_x$.

Define $V_y \subset \tilde{\mathcal{V}}/\pi$ by $\nu^{-1}(V_y) = \bigcup_{h \in \pi} hU_x$. Now the first hypothesis on U_x implies

$$0 \to Z \to \Gamma(U_x;\mathcal{O}(V^n)) \xrightarrow{e} \Gamma(U_x;\mathcal{O}^*(V^n)) \to 0$$

is an exact sequence of right $Z(\pi_x)$-modules. But then we have a commutative diagram

$$\begin{array}{ccccccc}
0 \to Z & \nearrow & \Gamma(U_x;\mathcal{O}(V^n)) & \xrightarrow{e} & \Gamma(U_x;\mathcal{O}^*(V^n)) & \searrow & 0 \\
 & \searrow & \uparrow & & \uparrow & \nearrow & \\
 & & C & \xrightarrow{\exp} & C^* & &
\end{array}$$

where C, C^* are trivial $Z(\pi_x)$-modules. Now $H^p(\pi_x;C) \simeq H^p(\pi_x;\Gamma(U_x,\mathcal{O}(V^n))) = 0$, $p > 0$ since π is finite,

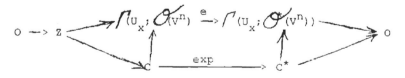

C and $\Gamma(U_x, \mathcal{O}(V^n))$ are vector spaces. Hence

$$H^p(\pi_x, \Gamma(U_x, \mathcal{O}^*(V^n)) \simeq H^{p+1}(\pi_x; Z) \simeq H^p(\pi_x; C)$$

for all $p > 0$. Incidentally $H^1(\pi_x; Z) = 0$ since π_x is finite and Z is a trivial $Z(\pi_x)$-module. If we apply (8.1) it follows that

$$H^p(\pi_x, C^*) \simeq H^p(\pi_x, \Gamma(U_x, \mathcal{O}^*(V^n))$$

$$H^p(\pi, \Gamma(\nu^{-1}(V_y), \mathcal{O}^*(V^n)), \; p > 0.$$

Since U_x is arbitrarily small the result follows. In addition, for $p = 0$, we see

$$0 \to \Gamma(\nu^{-1}(V_y); \mathcal{Z}(V^n))^{\pi} \to \Gamma(\nu^{-1}(V_y), \mathcal{O}(V^n))^{\pi}$$

$$\to \Gamma(\nu^{-1}(V_y), \mathcal{O}^*(V^n))^{\pi}$$

is also exact as $H^1(\pi; \Gamma(\nu^{-1}(V_y), \mathcal{Z}(V^n)) = 0$. Thus we also have a direct proof that

$0 \to \mathcal{Z}(V^n/\pi) \to \mathcal{O}(V^n/\pi) \xrightarrow{e} \mathcal{O}^*(V^n/\pi) \to 0$ is again exact.

Now the edge homomorphisms

$0 \to {}'E_2^{1,0} \to H^1(\pi; \mathcal{O}^*(V^n)) \to {}'E_2^{0,1}$ yield an exact sequence:

$$0 \to H^1(V^n/\pi; \mathcal{O}^*(V^n/\pi)) \to H^1(\pi; \mathcal{O}^*(V^n))$$

$$\to H^0(V^n/\pi; \mathcal{h}^1) \xrightarrow{d_2} H^2(V^n/\pi; \mathcal{O}^*(V^n/\pi)).$$

If $\nu(x) = y$ then $\mathcal{h}^1_y \simeq H^1(\pi_x; C^*) \simeq \mathrm{Hom}\,(\pi_x, C^*)$ as we just saw.

This suggests a particular interpretation of $H^1(\pi; \mathcal{O}^*(V^n)) \longrightarrow H^0(V^n/\pi; \mathcal{R}^1)$. To each $\xi \in H^1(\pi; \mathcal{O}^*(V^n))$ and x V^n we associate a characteristic homomorphism

$$R_x : \pi_x \longrightarrow C^*$$

as follows. Choose any holomorphic co-ordinate system with operators $r_{j,i}(h,x)$ representing ξ. If $x \in V_j$ then $R_x(h) = r_{j,j}(h,x)$ is a homomorphism of π_x into C^*. If $x \in V_i$ also then for $h \in \pi_x$

$$r_{j,j}(h,x)r_{j,i}(e,x) = r_{j,i}(e,x)r_{i,i}(h,x)$$

so the definition of R_x does not depend on the choice of V_j. If $r'_{j,i}(h,x)$ is holomorphically equivalent to $r_{j,i}(h,x)$ then for $h \in \pi_x$

$$r'_{j,j}(h,x)\lambda_j(x) = \lambda_j(x)r_{j,j}(h,x),$$

hence R_x only depends on ξ, not on how ξ is represented. Finally, suppose x is replaced by $\hat{h}x$, then $\pi_{\hat{h}x} = \hat{h}\pi_x\hat{h}^{-1}$ and

$$R_{\hat{h}x}(\hat{h}h\hat{h}^{-1}) = r_{j,j}(\hat{h}h\hat{h}^{-1},\hat{h}x) = r_{j,j}(\hat{h},x)r_{j,j}(h\hat{h}^{-1},\hat{h}x)$$

$$= r_{j,j}(\hat{h},x)r_{j,j}(h,x)r_{j,j}(\hat{h}^{-1},\hat{h}x).$$

But $1 = r_{j,j}(e,x) = r_{j,j}(\hat{h}^{-1}\hat{h},x) = r_{j,j}(\hat{h}^{-1},\hat{h}x)r_{j,j}(\hat{h},x)$ so that $R_{\hat{h}x}(\hat{h}^{-1}h\hat{h}) = R_x(h)$ for all $h \in \pi_x$. Clearly ξ lies in the kernel of $H^1(\pi; \mathcal{O}^*(V^n)) \longrightarrow H^0(V^n/\pi; \mathcal{R}^1)$ if and only if the characteristic homomorphism $R_x : \pi_x \longrightarrow C^*$ is trivial for all $x \in V^n$. Thus

(10.4) Theorem: An element of $H^1(\pi;\mathcal{O}^*(V^n))$ lies in the image of $H^1(V^n/\pi;\mathcal{O}^*(V^n/\pi))$ if and only if its characteristic homomorphism is trivial at each point of V^n.

The conclusion means that the element in question can be represented by $\{r_{j,i}(h,x)\}$ with $r_{j,j}(h,x) \equiv 1$, all h,x and j and $r_{j,i}(e,hx) \equiv r_{j,i}(e,x)$. We do not claim every representative has this form, only that at least one does.

Since the sequence

$$0 \to \mathcal{Z}(V^n/\pi) \to \mathcal{O}(V^n/\pi) \to \mathcal{O}^*(V^n/\pi) \to 0$$

is exact we may define $\mathcal{P}(V^n/\pi)$ to be the kernel of

$$c : H^1(V^n/\pi;\mathcal{O}^*(V^n/\pi)) \to H^2(V^n/\pi;Z).$$

Now from the 'E-spectral sequence for $H^*(\pi;\mathcal{O}(V^n))$ we see

$$H^p(V^n/\pi;\mathcal{O}(V^n/\pi)) \simeq H^p(\pi;\mathcal{O}(V^n)) \simeq H^p(V^n;\mathcal{O}(V^n))^{\pi}$$

since $\mathcal{h}^s = 0$, $s > 0$ in this case. Furthermore, $\mathcal{h}^1 \to V^n/\pi$ in the 'E-spectral sequence for $H^*(\pi;\mathcal{Z}(V^n))$ is also the zero sheaf, thus

$$H^1(V^n/\pi;Z) \simeq H^1(\pi;\mathcal{Z}(V^n)) \simeq H^1(V^n;Z)^{\pi}$$

Therefore, we conclude that

(10.5) Corollary: The monomorphism $H^1(V^n/\pi;\mathcal{O}^*(V^n/\pi)) \to H^1(\pi;\mathcal{O}^*(V^n))$ induces an isomorphism

$$\wp_{(V^n/\pi)} \simeq \wp_{(\pi,V^n)}.$$

By way of examples recall that by (4.3) if (π,V^n) has at least one stationary point, then

$$H^1(\pi;\mathcal{O}^*(V^n)) \simeq \mathrm{Hom}\,(\pi,C^*) \oplus H^1(V^n;\mathcal{O}^*(V^n))^{\pi}.$$

If in addition $H^1(V^n;Z) = 0$, then $H^1(V^n;\mathcal{O}^*(V^n))^{\pi} \subset H^2(V^n;Z)^{\pi}$ as the invariant elements in the kernel of $H^2(V^n;Z) \to h^{0,2}(V^n)$. The reader should apply this to an action of π on CP(n) with at least one fixed point. Another example would be the action of the symmetric group on a product of CP(n) with itself. Note especially that if n = 1 the quotient space of the symmetric group on $CP(1)^k$ is CP(k).

A more difficult example computationally is provided by an action of the cyclic group Z_n on the curve $V^1 \subset CP(2)$ defined by $\{[z_1,z_2,z_3] | z_1^n + z_2^n + z_3^n = 0\}$. With $\lambda = \exp(2\pi i/n)$ the action of Z_n is generated by $[z_1,z_2,z_3] \to [z_1,z_2,\lambda z_3]$. The quotient space V^1/Z_n is CP(1) with quotient map $[z_1,z_2,z_3] \to [z_1,z_2]$. Obviously $\wp(Z_n,V^1) = 0$ in view of (10.5), but $\wp(V^1)$ has complex dimension equal to genus $(V^1) = (n-1)(n-2)/2$, and $\wp(V^1)^{Z_n} = H^1(Z_n;H^1(V^1;Z))$. This is $\wp(V^n) \cap H^1(V^1;\mathcal{O}^*(V^1))^{Z_n}$. Then $H^1(Z_n;H^1(V^n;Z)) \subset H^1(V^1;\mathcal{O}^*(V^n))^{Z_n}$. The quotient naturally is isomorphic to a subgroup of $H^2(V^1;Z) = Z$. Is it the whole group or just the elements divisible by n?

Finally, suppose (Z_p,V^n) is a cyclic group of prime order acting with a connected non-empty fixed point set, then

$$H^1(Z_p;\mathcal{O}^*(V^n)) \cong \mathrm{Hom}\,(Z_p,C^*) \oplus H^1(V^n/Z_p;\mathcal{O}^*(V^n/Z_p)).$$

We define

$$H^1(Z_p;\mathcal{O}^*(V^n)) \longrightarrow \mathrm{Hom}\,(Z_p,C^*)$$

by choosing a fixed point and assigning to each element of $H^1(Z_p;\mathcal{O}^*(V^n))$ its characteristic homomorphism at this fixed point. If this characteristic homomorphism is trivial at this fixed point then, since the fixed point set of (Z_p,V^n) is connected, the characteristic homomorphism is trivial at every fixed point. At a non-fixed point, the characteristic homomorphism is obviously trivial since p is prime. Thus

$$0 \longrightarrow H^1(V^n/Z_p;\mathcal{O}^*(V^n/Z_p)) \longrightarrow H^1(Z_p;\mathcal{O}^*(V^n)) \longrightarrow \mathrm{Hom}\,(Z_p,C^*) \longrightarrow 0$$

exact. The reader may supply the splitting homomorphism. But this is the same splitting homomorphism that was used in (4.3), therefore we must also have

$$H^1(V^n/Z_p;\mathcal{O}^*(V^n/Z_p)) \cong H^1(V^n;\mathcal{O}^*(V^n))^{Z_p}.$$

(10.6) <u>Corollary</u>: <u>If (Z_p,V^n) has a non-empty connected fixed point set and if $H^1(V^n;Z) = 0$ then for every cohomology class of type (1,1), $c \in H^1(V^n;Z)^{Z_p}$</u>

$$\langle c^n, \sigma_{2n}\rangle = 0 \bmod p.$$

We see that

$$0 \longrightarrow H^1(V^n;\mathcal{O}^*(V^n))^{Z_p} \longrightarrow H^2(V^n;Z)^{Z_p} \longrightarrow h^{2,0}$$

is exact. So there is an element $\xi \in H^1(V^n;\mathcal{O}^*(V^n))^{Z_p}$ with $c_1(\xi) = c$. But ξ lies in the image of $H^1(V^n/Z_p;\mathcal{O}^*(V^n/Z_p)) \longrightarrow H^1(V^n;\mathcal{O}^*(V^n))$, hence there is a $\tilde{c} \in H^2(V^n/Z_p;Z)$ with $\nu^*(\tilde{c}) = c$, and hence $\nu^*(\tilde{c}^n) = c^n$. Since $\nu : H^{2n}(V^n/Z_p;Z) \longrightarrow H^{2n}(V^n;Z_p) \simeq Z$ has image pZ, the corollary follows. If the Kahler class is integral (i.e. V^n is algebraic) then the corollary applies to it.

11. Maps of prime period

Consider an action (Z_p, X), then $(\mathcal{Z}, Z_p) \to (Z_p, X)$ is the sheaf of germs of maps into Z with operators $(fh)(x) = f(hx)$. There is the fixed point set $F \subset X$. The use of $H^*(\pi;\mathcal{Z})$ in this case lies in the exact sequence

$$\ldots \to H^n(X/Z_p, F; Z) \to H^n(Z_p;\mathcal{Z}) \to (\sum_{i+j=n} H^i(Z_p; H^j(F;Z))$$
$$\to H^{n+1}(X/Z_p, F; Z) \to \ldots$$

where $H^j(F;Z)$ is a trivial $Z(Z_p)$ module.

In our framework this is seen as follows. There is the subsheaf $(\mathcal{Z}_F; Z_p) \subset (\mathcal{Z}, Z_p)$ consisting of the germs which vanish along F. Via the quotient map

$$\nu : X \to X/Z_p$$

we induce a sequence

$$0 \to (\mathcal{Z}^*_F, Z_p) \to (\mathcal{Z}^*, Z_p) \to (\mathcal{Z}^*(F), Z_p) \to 0$$

over X/Z_p where $\mathcal{Z}(F) = \mathcal{Z}/\mathcal{Z}_F$. We show immediately that Z_p acts trivially on $\mathcal{Z}^*(F)$. We apply (6.9) so that $H^n(Z_p; \mathcal{Z}^*(F)) \simeq \sum_{i+j=n} H^i(Z_p; H^j(X/Z_p; \mathcal{Z}(F))$. However $H^j(X/Z_p; \mathcal{Z}^*(F)) \simeq H^j(F;Z)$. To see $H^n(Z_p;\mathcal{Z}_F) \simeq H^n(X/Z_p, F; Z)$ just apply the 'E-spectral sequence and (8.4).

If $n > \dim X$, then $H^n(Z_p;\mathcal{Z}) \simeq \sum_{n=i+j} H^i(Z_p; H^j(F;Z))$. This, when combined with the "E-spectral sequence for $H(Z_p;\mathcal{Z})$, yields relations between $H^*(X)$ and $H^*(F)$. In actual practice, [3], a considerable simplification follows if at the outset Z is replaced by the field Z_p.

Chapter II

Orientation Preserving Involutions

1. The bordism group $A_n(2k)$

The basic object to consider is a pair $(\xi \to B^n, \theta)$ wherein $\xi \to B^n$ is an orthogonal 2k-plane bundle over a compact n-manifold and θ is an orientation on the Whitney sum $\xi \oplus \tau \to B^n$ where $\tau \to B^n$ is the tangent bundle. Let $-(\xi \to B^n, \theta) = (\xi \to B^n, -\theta)$. We identify $(\xi' \to W^n, \theta')$ with $(\xi \to B^n, \theta)$ if and only if there is an orthogonal bundle equivalence

$$\begin{array}{ccc} \xi' & \xrightarrow{\Phi} & \xi \\ \downarrow & & \downarrow \\ W^n & \xrightarrow{\varphi} & B^n \end{array}$$

for which

1) φ is a diffeomorphism
2) the induced bundle equivalence

$$\begin{array}{ccc} \xi' \oplus \tau' & \xrightarrow{\Phi \oplus d\varphi} & \xi \oplus \tau \\ \downarrow & & \downarrow \\ W^n & \xrightarrow{\varphi} & B^n \end{array}$$

preserves the orientation.

A boundary operator $\partial(\xi \to B^n, \theta) = (\xi \to \partial B^n, \partial_*(\theta))$

can be defined as follows. Along ∂B^n identify $\xi \oplus \tau$ with $\xi \oplus \tau' \oplus \eta$, where $\tau' \to \partial B^n$ is the tangent bundle to the boundary and $\eta \to \partial B^n$ is the (trivial) normal line bundle. Now $(\xi \oplus \tau') \oplus \eta \to \partial B^n$ inherits an orientation from that of $(\xi \oplus \tau)$, while, at each point of ∂B^n, η is given its orientation by the outward pointing unit normal vector. There is a unique orientation of $\xi \oplus \tau' \to \partial B^n$ compatible with those of η and of $(\xi \oplus \tau') \oplus \eta$, and this is $\partial \mathcal{O}$.

The bordism group $A_n(2k)$ can now be defined. If M^n and V^n are closed n-manifolds then $(\xi_o \to M^n, \mathcal{O}_o)$ is bordant to $(\xi_1 \to V^n, \mathcal{O}_1)$ if and only if there is a $(\xi \to B^{n+1}, \mathcal{O})$ for which $\partial(\xi \to B^{n+1}, \mathcal{O}) = (\xi_o \to M^n, \mathcal{O}_o) \sqcup -(\xi_1 \to V^n, \mathcal{O}_1)$. The symbol $\sqcup$ denotes disjoint union. Denote a bordism class by $[\xi \to M^n, \mathcal{O}]$, and the collection of all such bordism classes by $A_n(2k)$. As usual, an abelian group structure is imposed on $A_n(2k)$ by disjoint union. We agree that $A_n(o) = \Omega_n$, the cobordism group of closed oriented n-manifolds regarded as carrying "o-plane bundles."

Set $\mathcal{A}_m = \sum_{n+2k=m} A_n(2k)$ and $\mathcal{A} = \sum_o^\infty \mathcal{A}_m$. We can define in $\mathcal{A}$ the structure of a graded commutative algebra with unit over Ω, the oriented cobordism ring. We define the product $[\xi \to M^n, \mathcal{O}][\xi' \to V^m, \mathcal{O}']$ as follows. Form the external Whitney sum $(\xi \oplus \tau) \times (\xi' \oplus \tau') \to M^n \times V^m$. This is given the product orientation $\mathcal{O} \times \mathcal{O}'$. The canonical equivalence $(\xi \oplus \tau) \times (\xi' \oplus \tau') \simeq (\xi \times \xi') \oplus (\tau \times \tau')$ then induces the desired orientation. Since $A_n(o) = \Omega_n$, $\Omega \subset \mathcal{Q}$, hence $\mathcal{Q}$ is also an Ω-algebra with unit.

To compute $A_n(2k)$ we must show that it is naturally isomorphic to the Atiyah bordism group, [12], of the covering involution (T, BSO(2k)) over BO(2k) and then apply Rosenzweig's results, [12].

An element of BSO(2k) is a 2k-plane together with an orientation. The involution T is to reverse that orientation. Briefly the Atiyah group $A_n(T, BSO(2k))$ is defined as follows. The basic object is a pair $((t, \tilde{B}^n), \tilde{f})$ where $(t, \tilde{B}^n)$ is a fixed point free orientation reversing diffeomorphism of period 2 on a compact oriented manifold together with an equivariant map $\tilde{f} : (t, \tilde{B}^n) \to (T, BSO(2k))$. There is $\partial((t, \tilde{B}^n), \tilde{f}) = ((t, \partial\tilde{B}^n), \tilde{f}/\partial\tilde{B}^n)$ where $\partial\tilde{B}^n$ receives the induced orientation from that of $\tilde{B}^n$. The reader may go on to complete the definition of $A_n(T, BSO(2k))$.

We are given $(\xi \to B^n, \mathcal{O})$. Let $\tilde{B}^n$ be the set of all pairs (b, o), where $b \in B^n$ and o is an orientation of the fibre of ξ at b; that is, of the linear space ξ_b. There is an obvious fixed point free involution $t(b, o) = (b, -o)$ and a projection $\nu : \tilde{B}^n \to B^n$ given by $\nu(b, o) = b$. Topologize $\tilde{B}^n$ so that $\nu : \tilde{B}^n \to B^n$ is the principal Z_2-covering associated with the Whitney class $v_1(\xi) \in H^1(B^n; Z_2)$. Then $\tilde{B}^n$ receives a unique differential structure in which t is a diffeomorphism and for which $\nu : \tilde{B}^n \to B^n$ is a local diffeomorphism.

Now the induced bundle $\tilde{\xi} = \nu^{-1}(\xi)$ consists of all triples (b,o,v), where $v \in \xi_b$ and o is an orientation of ξ_b. Clearly $\tilde{\xi}$ is a canonically oriented 2k-plane bundle. Further, there is a bundle map $t^*(b,o,v) = (b,-o,v)$ covering t. This is an orientation reversing bundle involution because the orientation on $\tilde{\xi}$ induced by t^* is the reverse of the canonical orientation.

Next there is the bundle $\nu^{-1}(\tau) = \tilde{\tau} \to \tilde{B}^n$ induced by the tangent bundle to B^n. A point in $\tilde{\tau}$ is a triple (b,o,w) where $w \in \tau_b$ and o is an orientation of ξ_b. If we identify $\tilde{\tau}$ with the tangent bundle to $\tilde{B}^n$, then $dt(b,o,w) = (b,-o,w)$. We can just as well orient $\tilde{\tau} \to \tilde{B}^n$ because $\xi_b \oplus \tau_b$ is oriented and at (b,o) we can choose the orientation of $\tilde{\tau}_{(b,o)} = \tau_b$ which is compatible with o on ξ_b. Of course

$$\begin{array}{ccc} \tilde{\tau} & \xrightarrow{dt} & \tilde{\tau} \\ \downarrow & & \downarrow \\ \tilde{B}^n & \xrightarrow{t} & \tilde{B}^n \end{array}$$

is still orientation reversing. Finally, the product orientation on $\tilde{\xi} \oplus \tilde{\tau}$ is the same as the induced orientation from $\xi \oplus \tau$, and it is preserved by $t^* \oplus dt$.

The classifying map $f: B^n \to BO(2k)$ then lifts to an equivariant $\tilde{f} : (t,\tilde{B}^n) \to (T,BSO(2k))$, which classifies $\tilde{\xi}$. The process can be reversed. Suppose we are given a fixed point free orientation reversing involution on a compact oriented manifold together with an equivariant map $\tilde{f} : (t,\tilde{B}^n) \to (T,BSO(2k))$. The $\tilde{f}$ induces an oriented 2k-plane bundle $\widetilde{\xi} \longrightarrow \widetilde{B}^n$ with an orientation reversing involution $(t^*, \xi) \to (t, B^n)$. There is also the orientation reversing $(dt, \widetilde{\tau}) \longrightarrow (t, \widetilde{B}^n)$ and $(t^* \oplus dt, \widetilde{\xi} \oplus \widetilde{\tau})$ is orientation preserving. We can set $B^n = \widetilde{B}^n/t$, and orient the quotient bundle $\xi \oplus \tau = (\widetilde{\xi}/t^*) \oplus (\widetilde{\tau}/dt)$. This will lead the reader to see immediately that $A_n(2k) \simeq A_n(T, BSO(2k))$.

Let $M(\gamma)$ be the Thom space of the real line bundle $\gamma \longrightarrow BO(2k)$ associated with the double covering $BSO(2k) \longrightarrow BO(2k)$. According to Rosenzweig, [12], it follows now that

$$A_n(2k) \simeq \widetilde{\Omega}_{n+1}(M(\gamma))$$

where $\widetilde{\Omega}_{n+1}(M(\gamma))$ is the ordinary reduced oriented bordism group of the space $M(\gamma)$. For $k = o$, take $\gamma \longrightarrow \{pt = BO(o)\}$, then the Thom space is S^1 and surely $\Omega_n \simeq \widetilde{\Omega}_{n+1}(S^1)$.

2. Self-intersection

We continue to denote by $BO(2k)$ the classifying space for $O(2k)$, and by $MO(2k)$ we denote the Thom space of the universal $2k$-plane bundle. We shall define a homomorphism

$$S : A_n(2k) \longrightarrow \Omega_{n-2k}(BO(2k))$$

where $\Omega_*(BO(2k))$ is the ordinary bordism module of orthogonal $2k$-plane bundles over closed oriented manifolds.

Given $(\xi \longrightarrow M^n, \theta)$ there is a bundle map

$$\begin{array}{ccc} \xi & \xrightarrow{F} & \xi_{2k} \\ \downarrow & & \downarrow \\ M^n & \xrightarrow{f} & BO(2k). \end{array}$$

Now $BO(2k) \subset MO(2k)$, so by a small homotopy we can deform f into a map $g : M^n \xrightarrow{n} MO(2k)$ which is transverse regular on $BO(2k)$. Then

$$V^{n-2k} = g^{-1}(BO(2k))$$

is a closed regular submanifold of M^n. The normal bundle, denoted by $\hat{\xi} \longrightarrow V^{n-2k}$ is just the restriction of ξ. But $\xi \oplus \tau \mid V^{n-2k}$ is identified with $\hat{\xi} \oplus (\hat{\xi} \oplus \hat{\tau}) \longrightarrow V^{n-2k}$, where $\hat{\tau} \longrightarrow V^{n-2k}$ is the tangent bundle. Observe $\hat{\xi} \oplus (\hat{\xi} \oplus \hat{\tau}) = (\hat{\xi} \oplus \hat{\xi}) \oplus \hat{\tau}$ inherits an orientation from that of $\xi \oplus \tau \longrightarrow M^n$, while $\hat{\xi} \oplus \hat{\xi}$ is canonically oriented by the complex structure $(v,w) \longrightarrow (-w,v)$. Thus $\hat{\tau} \longrightarrow V^{n-2k}$ is oriented and $\hat{\xi} \longrightarrow V^{n-2k}$ represents an element in $\Omega_{n-2k}(BO(2k))$.

A standard transverse regularity argument shows that this construction defines a homomorphism
$S : A_n(2k) \to \Omega_{n-2k}(BO(2k))$.

To justify our calling S the self-intersection homomorphism we should note that V^{n-2k} is the set of zeroes of a "generic" cross-section of $\xi \to M^n$. In other words, any cross-section χ of $\xi \to M^n$ furnishes us with a map of M^n into the Thom space $M(\xi)$. The section is "generic" if this map is transverse regular on $M^n \subset M(\xi)$ as the 0-section. Let $\tilde{g} : M^n \to BO(2k)$ be the composition

$$M^n \xrightarrow{g} \xi_{2k} \xrightarrow{\pi} BO(2k)$$

where $\pi : \xi_{2k} \to BO(2k)$ is projection. Then $\tilde{g}$ is homotopic to $f : M^n \to BO(2k)$. Thus $\tilde{g}^{-1}(\xi_{2k})$ is equivalent to ξ, but $x \to (x, g(x))$ is a "generic" cross-section of this induced bundle whose set of zeroes is exactly V^{n-2k}.

Let us have an example then. For each $n \geq 0$ consider the holomorphic line bundle $\eta_n \to CP(n)$ over complex projective space defined as follows. Let $W = C^{n+1} \setminus \{0\}$ and define the action of C^* on $C \times W$ by $t(\mathfrak{z}; z_1, \ldots, z_{n+1}) = (t\mathfrak{z}; tz_1, \ldots, tz_{n+1})$. Then $\eta_n = (C \times W)/C^* \to CP(n)$. If we give $CP(n)$ the orientation arising from its analytic structure then we obtain $[\eta_n \to CP(n)]$ in $\Omega_{2n}(BO(2))$. If $\eta_{n+1} \oplus \tau \to CP(n+1)$ is oriented by the complex structure on this Whitney sum then we obtain $[\eta_{n+1} \to CP(n+1), \varnothing] \in A_{2n+2}(2)$. We assert that for

$n \geq 0$

$$S[\eta_{n+1} \to CP(n+1), \mathcal{O}] = -[\eta_n \to CP(n)].$$

The generic cross-section of $\eta_{n+1} \to CP(n+1)$ is

$$\chi[z_1, \ldots, z_{n+2}] = ((z_1, z_1, \ldots, z_{n+2}))$$

and the set of zeroes of this holomorphic section is

$$CP(n) = \{[0, z_2, \ldots, z_{n+2}]\} \subset CP(n+1)$$

and $\hat{\eta}_{n+1} = \eta_n \to CP(n)$. Now $\eta_{n+1} \oplus \tau \to CP(n+1)$ had the orientation of the complex structure, so that $\eta_n \oplus \eta_n \oplus \hat{\tau} \to CP(n)$ has the "correct" orientation, but $\eta_n \oplus \eta_n$ is oriented by the complex structure $(v,w) \to (-w,v)$. A moments reflection shows that from this $\hat{\tau} \to CP(n)$ receives the orientation opposite its customary one, hence the minus sign.

A small, well known, comment on orientations is appropriate here. Let $(V, \mathcal{O})$ be an oriented real vector space of dimension n. The direct sum $V \oplus V$ can be given either the product orientation $\mathcal{O} \times \mathcal{O}$ or the orientation $\mathcal{O}_c$ arising from the complex structure $(v,w) \to (-w,v)$. In every case

$$\mathcal{O} \times \mathcal{O} = (-1)^{\frac{n(n-1)}{2}} \mathcal{O}_c.$$

With this in mind, we might ask how the self-intersection is related to the fact that $A_n(2k) \simeq A_n(T, BSO(2k))$. Given $(\xi \to M^n, \mathcal{O})$ we took a generic cross-section χ of ξ and found V^{n-2k} to be the set of zeroes. But χ will induce a generic cross-section of $\tilde{\xi} \to \tilde{M}^n$ whose set of zeroes, $\tilde{V}^{n-2k}$, is t-invariant and double covers V^{n-2k}. There is an ambiguity in orienting $\tilde{V}^{n-2k}$, however. We might use $(\tilde{\xi}^\wedge \oplus \tilde{\xi}^\wedge) \oplus \tilde{\tau}^\wedge$ just as before giving $\tilde{V}^{n-2k}$ an orientation $\mathcal{O}_c$ in which $\tilde{V}^{n-2k} \to V^{n-2k}$ has degree +2. However, $\tilde{\xi}$, $\tilde{\tau}$ are both oriented and $\tilde{\tau}/\tilde{V}^{n-2k} = \tilde{\xi}^\wedge \oplus \tilde{\tau}^\wedge$ will also orient $\tilde{V}^{n-2k}$, say with orientation $\tilde{\mathcal{O}}$. This amounts to giving $\tilde{\xi}^\wedge \oplus \tilde{\xi}^\wedge$ the product orientation so $\tilde{\mathcal{O}} = (-1)^k \mathcal{O}_c$. Now $(\tilde{V}^{n-2k}, \tilde{\mathcal{O}})$ is the oriented submanifold of M^n dual to the Euler class $X(\tilde{\xi}) \in H^{2k}(\tilde{M}^{2n};Z)$ of the oriented bundle $\tilde{\xi} \to \tilde{M}^n$.

We can use this commentary to produce another example which may be illuminating. Given a closed manifold, M^{2k}, let $\xi \to M^{2k}$ be the tangent bundle, then $\xi \oplus \tau \to M^{2k}$ is oriented by the complex structure $(v,w) = (-w,v)$. In this case $\tilde{\xi} \to \tilde{M}^{2k}$ is the canonically oriented tangent bundle of the orientation double covering of M^n, but we must pause to consider the orientation of $\tilde{\tau} \to \tilde{M}^n$. It arises from the orientation of $\tilde{\xi}$ together with the complex structure on $\tilde{\xi} \oplus \tilde{\tau}$. We can write $\tilde{\tau} = (-1)^k \tilde{\xi}$ as oriented bundles since the identity map is an orthogonal bundle equivalence between $\tilde{\tau}$ and $\tilde{\xi}$. Thus $\tilde{V}^o$ with the $\mathcal{O}_c$ orientation is dual to $(-1)^k X(\tilde{\xi}) = (-1)^{2k} X(\tilde{\tau}) = \chi(\tilde{M}^{2k})$. Now $\chi(M^{2k}) = 1/2\, \chi(\tilde{M}^{2k}) = 1/2 \langle X(\tilde{\tau}), \tilde{\sigma}_{2k}\rangle$. Since $S[\xi \to M^{2k}, \mathcal{O}] = [V^o] \in \Omega_o(BO(o)) = \Omega_o \simeq Z$ we can write

$$S[\xi \to M^{2k}] = \chi(M^{2k}),$$

because $\chi(\tilde{M}^{2k}) = \langle X(\tilde{\tau}), \tilde{\sigma}_{2k}\rangle$ is the number of points in $\tilde{V}^o$ counted with sign in the σ_c orientation, and thus is just twice the number of points in V^o also counted with sign.

Let us now investigate the multiplicative properties of the self-intersection homomorphism. There is

$$BO(2k) \times BO(2j) \to BO(2(k+j))$$

inducing

$$MO(2k) \wedge MO(2j) \to MO(2(k+j)),$$

and this second map is already transverse regular on $BO(2(k+j))$. Thus if $g : M^n \to MO(2k)$, $g' : L^m \to MO(2j)$ are suitably transverse regular, then

$$g \times g' : M^n \times L^m \to MO(2k) \wedge MO(2j) \to MO(2(k+j))$$

is still transverse regular on $BO(2(k+j))$ and $(g \times g')^{-1}(BO(2(k+j))) = V^{n-2k} \times W^{m-2j}$, with $\hat{\xi} \times \hat{\xi}' \to V^{n-2k} \times W^{m-2j}$ as normal bundle. We must show that $\hat{\tau} \times \hat{\tau}' \to V^{n-2k} \times W^{m-2j}$ receives the product orientation. First the bundle $(\xi \oplus \tau) \times (\xi' \oplus \tau')$ receives the product orientation and restricts into

$$\alpha = ((\hat{\xi} \oplus \hat{\xi}) \oplus \hat{\tau}) \times ((\hat{\xi}' \oplus \hat{\xi}') \oplus \hat{\tau}') \to V^{n-2k} \times W^{m-2j}.$$

so the inherited orientation of α agrees with the product of the orientations on $(\hat{\xi} \oplus \hat{\xi})$, $\hat{\tau}$, $(\hat{\xi}' \oplus \hat{\xi}')$, $\hat{\tau}'$. Now

$$\beta = ((\hat{\xi} \oplus \hat{\xi}) \times (\hat{\xi}' \oplus \hat{\xi}')) \oplus (\hat{\tau} \times \hat{\tau}')$$

is also oriented and the natural equivalence $\alpha \simeq \beta$ is orientation preserving. Next, there is the complex structure on $(\hat{\xi} \times \hat{\xi}') \oplus (\hat{\xi} \times \hat{\xi}')$ which induces an orientation on $\gamma = (\hat{\xi} \times \hat{\xi}') \oplus (\hat{\xi} \times \hat{\xi}') \oplus (\hat{\tau} \times \hat{\tau}')$. Since the obvious isomorphism $\beta \simeq \gamma$ is the sum of a complex linear isomorphism with the identity, it too preserves orientation. Thus $\alpha \simeq \gamma$ by an orientation preserving equivalence. Now γ is the restriction of $(\xi \times \xi') \oplus (\tau \times \tau')$ which was oriented by equivalence to $(\xi \oplus \tau) \times (\xi' \oplus \tau')$ and $\alpha \simeq \gamma$ is the restriction of this equivalence. Thus the inherited orientation of γ is the same as the product orientation arising from $\hat{\tau}$, $\hat{\tau}'$ and the complex bundle $(\hat{\xi} \times \hat{\xi}') \oplus (\hat{\xi} \times \hat{\xi}')$, therefore $V^{n-2k} \times W^{m-2j}$ does receive the product orientation.

This suggests that we put $\mathcal{M}_m = \sum_{p+4q=m} \Omega_p(BO(2q))$ and $\mathcal{M} = \sum_o^\infty \mathcal{M}_m$. Via the external Whitney sum we define in $\mathcal{M}$ the structure of a graded commutative algebra with unit over Ω. Following the agreement $A_n(o) = \Omega_n = \Omega_n(BO(o))$ we let $S : A_n(o) \to \Omega_n(BO(o))$ be the identity. The grading on $\mathcal{M}$ was chosen just so that $S : \mathcal{A} \to \mathcal{M}$ will be a unit preserving homomorphism of graded algebras having degree 0.

3. The structure of $\mathcal{A}$

In this section we shall prove that $\mathcal{A}$/Tor is a graded polynomial ring over Ω/Tor. We begin with two new remarks about general bordism theory which are natural generalizations of results in [6].

(3.1) Lemma: If $H_*(X;Z)$ has no odd torsion then $\Omega_*(X) \otimes Z(1/2)$ is a free graded module over $\Omega \otimes Z(1/2)$.

Under this hypothesis, the bordism spectral sequence for X collapses, [6, Ch. II], and in particular the Thom homomorphism $\mu : \Omega_*(X) \to H_*(X;Z)$ is onto. If rank $H_n(X;Z) \neq 0$ choose elements $Y_{n,i} \in \Omega_n(X)$ where $1 \leq i \leq$ rank $H_n(X;Z)$ so that the homology classes $\mu(Y_{n,i})$ are an additive base for the free part of $H_n(X;Z)$. Let $\mathcal{Y}\Omega_*(X)$ be the submodule over Ω generated by all the $Y_{n,i}$. There is a filtration $\Omega_n(X) = J_{n,o} \supset \dots \supset J_{o,n} \supset 0$ with

$$J_{p,n-p} \mid J_{p-1,n-p+1} \simeq E^{\infty}_{p,n-p} \simeq E^2_{p,n-p}$$

$$\simeq H_p(X;\Omega_{n-q}).$$

We show by induction on p that if $\alpha \in J_{p,n-p}$ then $2^j\alpha$ for some $j \geq 0$. Suppose this has been shown for p-1, then α determines $\hat{\alpha} \in J_{p,n-p} | J_{p-1,n-p+1} \simeq H_p(X;\Omega_{n-p})$. But there is an exact sequence

$$0 \to H_p(X) \otimes \Omega_{n-p} \to H_p(X;\Omega_{n-p}) \to \mathrm{Tor}(H_{p-1}(X),\Omega_{n-p}) \to 0.$$

Certainly there is a $2^j\alpha$ and elements $x_i^{n-p} \in \Omega_{n-p}$ such that $2^j\alpha - (\sum Y_{p,i} x_i^{n-p}) \in J_{p-1,n-p+1}$ so the induction can proceed. Note that $J_{o,n} \simeq H(X; \Omega_n) \simeq H_o(X) \otimes \Omega_n$. The reader may show the $Y_{n,i}$ are linearly independent in $\Omega_*(X) \otimes Z(1/2)$.

Let us briefly recall the Pontrjagin numbers of a map of a closed manifold $f : M^n \to X$, [6,Ch.II]. For each finitely non-zero sequence $w = (o \leq i_1 \leq \ldots \leq i_j \leq \ldots)$ let $p(w) = p_1^{i_1} \ldots p_j^{i_1} \ldots$ be the product of the Pontrjagin classes of M^n. For each pair (w,c) where $c \in H^*(X;Z)$ define $P(w,c)[M^n,f] = \langle f^*(c)p(w), \sigma_n \rangle \in Z$. This is an invariant of the bordism class and determines $[M^n,f]$ mod torsion.

Consider especially $f : M_i^n \to X$ representing $Y_{n,i}$. If V^{4m} is a closed manifold we can compute some of the Pontrjagin numbers of $[M_i^n \times V^{4m},g]$ where $g(x,y) = f(x)$. Choose $c_{n,i} \in H^n(X;Z)$ with $\langle c_{n,i}, \mu(Y_{n,j}) \rangle = \delta_{i,j}$. If $c \in H_p(X;Z)$, $p > n$, then $P(w,c)[M_i^n \times V^{4m},g] = 0$ while

$$P(w,c_{n,j})[M_i^n \times V^{4m},g] = \delta_{i,j} \langle p(w), \sigma(V^{4m}) \rangle \in Z.$$

Both remarks follow upon application of the Whitney sum theorem to the computation of the Pontrjagin classes of $M_i^n \times V^{4m}$ together with appropriate dimensional considerations. This type of consideration was used in [6,Ch.II]. This means the Pontrjagin numbers of $[M_i^n \times V^{4m},g]$ determines the Pontrjagin numbers of $[V^{4m}]$.

(3.2) Theorem: If $H_*(X;Z)$ has no odd torsion, then $\Omega_*(X)/\mathrm{Tor}$ is a free graded Ω/Tor module.

Consider a

$$\beta = \sum_{n+4m=k} \left(\sum_i Y_{n,i}[V_i^{4m}]\right) \in \Omega_k(X)$$

The range on each i is $1 \leq i \leq \operatorname{rank} H_n(X;Z)$. We must show that if all the generalized Pontrjagin numbers of β are divisible by 2, then, modulo an element of order 2, each coefficient $[V_i^{4m}] \in 2\Omega_{4m}$. This is equivalent to showing all Pontrjagin numbers of $[V_i^{4m}]$ are divisible by 2. We proceed inductively. Fix n_o and assume the divisibility has been established for the coefficients $[V_i^n]$, all $n > n_o$. Without loss of generality we may then assume that in fact no term with $n > n_o$ appears. Fix i_o and consider $P(w,c_{n_o,i_o})\beta = \sum (\sum P(w,c_{n_o,i_o})[M_i^n \times V_i^{4m},g])$ with $n \leq n_o$. But all the contributions are 0 except

$$P(w,c_{n_o,i_o})\left[M_{i_o}^{n_o} \times V_{i_o}^{4m_o},g\right] = pw([V_{i_o}^{4m_o}]).$$

Since all Pontrjagin numbers of β are divisible by 2, so are all Pontrjagin numbers of $[V_{i_o}^{4m_o}]$. Thus the induction may proceed. It begins with $n = k$ of course.

Now if $2^j\alpha + \gamma = \beta \in \mathcal{Y}$ where γ is a torsion element, then we apply the foregoing to find a $\beta' \in \mathcal{Y}$ with $2(\beta - 2\beta') = 0$. Thus $2(\beta' - 2^{j-1}\alpha) - \gamma = 2\beta' - \beta$, hence $\beta' - 2\alpha^{j-1} = \gamma'$ is a torsion class and $2^{j-1}\alpha + \gamma' = \beta' \in \mathcal{Y}$. Inductively, therefore, we can, for any $\alpha \in \Omega_k(X)$, find a

$\beta \in \mathcal{Y}$ for which $\alpha - \beta$ is a torsion class. Thus the $Y_{n,i}$ generate $\Omega_*(X)/\mathrm{Tor}$ as an Ω/Tor-module. It is not difficult to verify the linear independence over Ω/Tor.

We want to apply (3.2) to $\Omega_*(BO(2k))$. This is possible for $H_*(BO(2k);Z)$ has no odd torsion and no elements of order 4. We have to find generators for the free part of $H_*(BO(2k);Z)$. For a start, the free part of $H^*(BO(2k);Z)$ is the polynomial ring generated by the Pontrjagin classes $p_1,\ldots,p_k$. Identify these with the elementary symmetric functions in the variables $t_1^2,\ldots,t_k^2$. To each $w = (0 \leq i_1 \leq \ldots \leq i_k)$ associate the symmetric function $\sum t_1^{2i_1},\ldots,t_k^{2i_k}$ which may be regarded as a polynomial in the Pontrjagin classes. Also to w associate the external Whitney sum of line bundles

$$Y(w) = [\eta_{2i_1} \times \ldots \times \eta_{2i_k} \to CP(2i_1) \times \ldots \times CP(2i_k)]$$

in $\Omega_{4(i_1+\ldots+i_k)}(BO(2k))$. If

$$f : CP(2i_1) \times \ldots \times CP(2i_k) \to BO(2k)$$

be the classifying map, then $\langle f^*(\sum t_1^{2i_1}\ldots t_k^{2i_k}), \sigma \rangle = 1$, where σ is the fundamental class of the product of projective spaces. Thus by (3.2) the Y(w) are a basis of $\Omega_*(BO(2k))/\mathrm{Tor}$ as a module over Ω/Tor. Recalling the definition of $\mathcal{M}$ we introduce $x_{4(p+1)} = [\eta_{2p} \to CP(2p)] \in \mathcal{M}_{4(p+1)}$ for all $p \geq 0$.

(3.3) Lemma: *As a graded algebra over Ω/Tor, $\mathcal{M}/\mathrm{Tor}$ is a polynomial algebra generated by the $[\eta_{2p} \to CP(2p)]$ for $p \geq 0$.*

We repeat that $\mathcal{M}$ contains no elements of order 4.

We return to $A_n(2k) \simeq \tilde{\Omega}_n(M(\gamma))$. Since $H_*(M(\gamma);Z)$ has no odd torsion and no elements of order 4, we can determine $\mathcal{Q} \otimes Z(1/2)$. We compute $\tilde{H}^*(M(\gamma);Z(1/2))$ as follows. Let $D(\gamma)$, $S(\gamma)$ respectively be the 1-cell and the 0-sphere bundles of $\gamma \to BO(2k)$. Then

$\tilde{H}^*(M(\gamma);Z(1/2)) \simeq H^*(D(\gamma), S(\gamma);Z(1/2))$ and

$H^*(D(\gamma), S(\gamma);Z(1/2)) \simeq H^*(BO(2k);Z(1/2)) \simeq Z(1/2)[p_1,\dots,p_k]$

$H^*(S(\gamma);Z(1/2)) \simeq H^*(BSO(2k);Z(1/2)) \simeq Z(1/2)[p_1,\dots,p_{k-1},X_{2k}]$

where X_{2k} is the universal Euler class and $i^*(p_k) = (X_{2k})^2$.

Thus $0 \to Z(1/2)[p_1,\dots,p_k] \xrightarrow{i^*} Z(1/2)[p_1,\dots,p_{k-1},X_{2k}] \xrightarrow{\delta} \tilde{H}^*(M(\gamma);Z(1/2)) \to 0$ is exact. It follows that $\tilde{H}^{n+1}(M(\gamma);Z(1/2)) = 0$ if $n-2k \neq 0 \bmod 4$. If $n-2k = 4j$ then rank $\tilde{H}^{n+1}(M(\gamma);Z(1/2))$ is the number of partitions $(0 \leq i_1 \leq \dots \leq i_k)$ of j. To each such partition we may correspond $\delta^*(p_1^{i_1}\dots p_{k-1}^{i_{k-1}}(X_{2k})^{2i_k}X_{2k}) \in \tilde{H}^{n+1}(M(\gamma);Z(1/2))$.

Now observe that additively

$$\tilde{H}^{n+1}(M(\gamma);Z(1/2)) \simeq H^{n-2k}(BO(2k);Z(1/2))$$

By (3.1) we have

$$\tilde{\Omega}_*(M(\gamma)) \otimes Z(1/2) \simeq H_*(M(\gamma);Z) \otimes (\Omega \otimes Z(1/2))$$

and so it follows that

$$\tilde{\Omega}_*(M(\gamma)) \otimes Z(1/2) \simeq A_*(2k) \otimes Z(1/2) \simeq \Omega_*(BO(2k)) \otimes Z(1/2).$$

Summing over 2k we conclude that an $\Omega \otimes Z(1/2)$-modules

$$\mathcal{O} \otimes Z(1/2) \simeq \mathcal{M} \otimes Z(1/2).$$

(3.4) <u>Theorem</u>: <u>The self-intersection induces an isomorphism</u>

$$S : \mathcal{O}/\mathrm{Tor} \simeq \mathcal{M}/\mathrm{Tor}$$

<u>of graded algebras</u>.

Recall from section 2 that we showed $S\,[\eta_{2p+1} \to CP(2p+1), \sigma] = -[\eta_{2p} \to CP(2p)]$ so S is onto mod torsion. Since $\mathcal{O} \otimes Z(1/2) \simeq \mathcal{M} \otimes Z(1/2)$ as modules the kernel of S consists of torsion only.

(3.5) <u>Corollary</u>: <u>As a graded algebra over</u> Ω/Tor, $\mathcal{O}/\mathrm{Tor}$ <u>is a polynomial ring on</u> $[\eta_{2p+1} \to CP(2p+1), \sigma]$, $p \geq 0$.

(3.6) <u>Corollary</u>: <u>Every element in the kernel of</u> S <u>has order</u> 2.

There are no elements of order 4 in $\mathcal{O}$.

4. The ring $\mathcal{O}_*(Z_2)$

We define $\mathcal{O}_m(Z_2)$ to be the bordism group of orientation preserving involutions on closed oriented m-manifolds, [1Z] . Give to $\mathcal{O}_*(Z_2) = \sum \mathcal{O}_m(Z_2)$ the structure of a graded commutative algebra with unit over Ω, where

$$[T, M^m][T', V^n] = [T \times T', M^m \times V^n].$$

We embed Ω into $\mathcal{O}_*(Z_2)$ by assigning to each closed oriented manifold the trivial involution.

Rosenzweig, [12], introduced an exact triangle

$$\begin{array}{ccc} \mathcal{A} & \xrightarrow{\partial} & \Omega_*(Z_2) \\ {\scriptstyle j_*}\nwarrow & & \swarrow{\scriptstyle i_*} \\ & \mathcal{O}_*(Z_2) & \end{array}$$

where $\Omega_*(Z_2)$ is the bordism module of fixed point free orientation preserving involutions on closed oriented manifolds. We define $i_* : \Omega_*(Z_2) \to \mathcal{O}_*(Z_2)$ by disregarding the freeness. Now $\mathcal{O}_*(Z_2) \to \mathcal{A}$ is described as follows. Given (T, M^n) then for each $0 \leq n \leq m$ let $F^n \subset M^m$ be the union of the n-dimensional components of the fixed point set. Let $\zeta \to F^n$ be the normal bundle. Since T preserves orientation at each fixed point $m-n = 2k$ if $F^n \neq \emptyset$. Finally, the orientation of M^n defines an orientation on $\zeta \oplus \tau \to F^n$. For each n, $F^n \neq \emptyset$ we have $[\zeta \to F^n, \mathcal{O}]$ in $A_n(2k)$, and

$[T, M^m] \to \sum [\xi \to F^n, \theta] \in \mathcal{A}_m$. This defines a homomorphism of algebras $\mathcal{O}_*(Z_2) \to \mathcal{A}$. Now given $(\xi \to V^n, \theta)$ we introduce the bundle involution $(A, S(\xi))$ on the (2k-1)-sphere bundle associated with ξ. This is the fibre preserving fixed point free involution which, on each fibre, agrees with the antipodal map. The orientation on $\xi \oplus \tau \to V^n$ produces an orientation on S( ) which is preserved by A so we put

$$\partial_* [\xi \to V^n, \theta] = [A, S(\xi)] \in \Omega_{n+2k-1}(Z_2).$$

As we said, this results in an exact triangle. Rosenzweig proved, [/Z], that the kernel of $i_* : \Omega_*(Z_2) \to \mathcal{O}_*(Z_2)$ consists of precisely all the 2-torsion. Now $\Omega_*(Z_2) \simeq \Omega \oplus \tilde{\Omega}_*(Z_2)$ so that we have

$$0 \longrightarrow \Omega/\mathrm{Tor} \longrightarrow \mathcal{O}_*(Z_2) \xrightarrow{j_*} \mathcal{A} \xrightarrow{\partial_*} \mathrm{Tor}\ \Omega_*(Z_2) \longrightarrow 0.$$

If $[Z_2, Z_2] \in \mathcal{O}_o(Z_2)$ is the action of Z_2 on itself by translation then $\Omega/\mathrm{Tor} \to \mathcal{O}_*(Z_2)$ is

$$[M^{4p}] \to [Z_2, Z_2][M^{4p}].$$

The image of this homomorphism is the ideal generated by $[Z_2, Z_2]$ and it will be denoted by $J \subset \mathcal{O}_*(Z_2)$. It is the kernel of $j_* : \mathcal{O}_*(Z_2) \to \mathcal{A}$.

For p > 0 introduce the $[A, CP(2p)]$ in $\mathcal{O}_{4p}(Z_2)$ where $A[z_1, \dots, z_{2p+1}] = [-z_1, z_2, \dots, z_{2p+1}]$.

(4.1) Theorem: As an algebra over Ω/Tor, $(\mathcal{O}_*(Z_2)/J)$/Tor is a polynomial algebra, with generators $[A, CP(2p)]$, $p > 0$.

The fixed point set of $(A, CP(2p+2))$ is the disjoint union of a point and $CP(2p+1) \subset CP(2p+2)$ as $\{[0, z_2, \ldots, z_{2p+3}]\}$ with normal bundle $\eta_{2p+1} \to CP(2p+1)$. When we apply the self-intersection we are left with $-[\eta_{2p} \to CP(2p)] = -x_{4(p+1)} \in \mathcal{M}_{4(p+1)}$. Thus (4.1) follows from the structure of $\mathcal{M}$/Tor and the fact that $\mathcal{O}_*(Z_2)/J$ is embedded in $\mathcal{a}$. We shall denote by $\mathcal{S}: \mathcal{O}_*(Z_2) \to \mathcal{M}$ the composite homomorphism

$$\mathcal{O}_*(Z_2) \xrightarrow{j_*} \mathcal{a} \xrightarrow{S} \mathcal{M}.$$

(4.2) Corollary: If $2\mathcal{S}[T, M^{4p}] = 0$ then $2[T, M^{4p}] = [Z_2, Z_2][M^{4p}]$ in $\mathcal{O}_*(Z_2)$.

The hypothesis implies $j_*[T, M^{4p}]$ has order 2 in $\mathcal{a}_{4p}$, hence there is $[Z_2, Z_2][X^{4p}] = 2[T, M^{4p}]$ for some $[X^{4p}]$. Ignoring the involutions we see that $2([X^{4p}] - [M^{4p}]) = 0$ so that $[Z_2, Z_2][X^{4p}] = [Z_2, Z_2][M^{4p}]$. From (4.2) and elementary dimensional considerations we derive

(4.3) Corollary: If (T, M^{4p}) is an orientation preserving involution on a closed manifold for which every component of the fixed point set has dimension less than 2p, then

$$2[T, M^{4p}] = [Z_2, Z_2][M^{4p}]$$

in $\mathcal{O}_{4p}(Z_2)$.

We can use (4.2) to answer the crucial question of identifying the elements in $2\mathcal{O}_*(Z_2)\cap J$. We shall have to examine again some involutions which were used elsewhere, [8].

For each pair of non-negative integers (n,k) we shall construct a holomorphic transformation group $(C^*, V(n,k))$ on a closed complex analytic manifold of real dimension $2(n+k)$. With k fixed we induct on n as follows. Let $V(o,k) = CP(k)$ and define $(C^*, V(o,k))$ by $t([\lambda_1,\ldots,\lambda_{k+1}]) = [\lambda_1,\ldots,\lambda_k,t\lambda_{k+1}]$. Suppose $(C^*, V(n,k))$ has been inductively defined, then $V(n+1,k)$ is the quotient of $(C^*, V(n,k)\times W)$ where $W = C^2\setminus\{o,o\}$ at $t(x,z,w) = (tx,tz,tw)$. Denoting by $((x,z,w))$ a point in $V(n+1,k)$, we define $(C^*, V(n+1,k))$ by $t((x,z,w)) = ((x,z,tw))$. We may fibre $V(n+1,k)$ over $CP(1)$ with fibre $V(n,k)$ by the projection $((x,z,w)) \longrightarrow [z,w]$. We are especially interested in $[T, V(n+1,k)] \in \mathcal{O}_{2(n+k+1)}(Z_2)$ where $T((x,z,w)) = ((x,z,-w))$. The fixed point set $F(n+1,k) \subset V(n+1,k)$ falls naturally into two parts. There is $V(n,k) \subset V(n+1,k)$ as $x \longrightarrow ((x,1,o))$, and there is $F(n,k) \subset V(n+1,k)$ as $x \longrightarrow ((x,o,1))$. We have $F(n+1,k) = F(n,k) \sqcup V(n,k)$. Furthermore the normal bundle to $V(n,k) \subset V(n+1,k)$ is an analytically trivial line bundle. In fact $V(n,k)\times C$ is mapped onto this normal bundle by $(x,w) \longrightarrow ((x,1,w))$. To understand the normal bundle to $F(n,k) \subset V(n+1,k)$ we should show it is the sum of the normal bundle of $F(n,k) \subset V(n,k)$ with a trivial line bundle. But if $V(n,k)$ is embedded in $V(n+1,k)$ by $x \longrightarrow ((x,o,1))$ then $F(n,k) \subset V(n,k) \subset V(n+1,k)$ and $V(n,k) \subset V(n+1,k)$ has a trivial normal bundle too. Hence the normal bundle to $F(n,k) \subset V(n+1,k)$ is the Whitney sum of the normal bundle of $F(n,k) \subset V(n,k)$ with a trivial line bundle as required. Inductively

$$F(n+1,k) = \Big(\coprod_{o<j<n+1} V(j,k)\Big) \coprod F(o,k).$$

Furthermore, the normal to this fixed point set admits a non-vanishing cross-section on every component.

(4.4) Lemma: *If* $n > 0$ *then*

$$2\,[T,V(n,k)] = [Z_2,Z_2][V(n,k)]$$

in $\mathcal{O}_{2(n+k)}(Z_2)$.

It is evident that $\mathcal{S}[T,V(n,k)] = 0$ if $n > 0$, so we apply (4.2). As $V(n,k)$ is fibred over $CP(1)$ for $n > 0$ we also have index $[V(n,k)] = 0$.

A set of generators for $\Omega/\text{Tor} = Z[x_4, x_8, \ldots, x_{4p}, \ldots]$ can be described by $x_4 = [CP(2)]$ and for $p > 1$, x_{4p} = integral linear combination of $[V(n,k)]$ with $n+k = 2p$, $n > 0$, [8]. But otherwise, we can choose generators of Ω/Tor such that for $p > 1$, $[Z_2,Z_2]x_{4p} \in 2\,\mathcal{O}_{4p}(Z_2)$.

For any $[M^{4p}]$ consider $[M^{4p}] - (\text{index}\,[M^{4p}])[CP(2)]^p$. This difference has index 0 so that it lies in the ideal generated by $\{x_{4p}\}_2$. Using the fact that $\Omega/\text{Tor} \to \mathcal{O}_*(Z_2)$ is an Ω-module homomorphism we have $[Z_2,Z_2]([M^{4p}] - (\text{ind}\,[M^{4p}])[CP(2)])^p \in 2\,\mathcal{O}_{4p}(Z_2)$. Furthermore

$[Z_2, Z_2]((\mathrm{ind}[M^{4p}])[CP(2)]^p) \in 2\mathcal{O}_{4p}(Z_2)$ if index $[M^{4p}] = 0 \bmod 2$.

(4.5) <u>Theorem</u>: <u>The necessary and sufficient condition that</u> $[Z_2, Z_2][M^{4p}] \in 2\mathcal{O}_*(Z_2)$ is

$$\text{index}\,[M^{4p}] = 0 \ \underline{\text{mod}}\ 2.$$

The reader will note of course that we have established the sufficiency and the necessity seems plausible, but we must postpone the proof of this because we are missing an important invariant which will be developed in the next section.

We note that Rosenzweig, [12], used his exact triangle to show $\mathcal{O}_*(Z_2)$ has no odd torsion and no elements of order 4. Now $J \cap \mathrm{Tor}\,\mathcal{O}_*(Z_2) = \{0\}$ so that we can regard $\mathrm{Tor}\,\mathcal{O}_*(Z_2)$ as contained in $\mathrm{Tor}(\mathcal{O}_*(Z_2)/J)$. Since $\mathcal{O}_*(Z\ )/J \subset \mathfrak{A}$, $\mathrm{Tor}(\mathcal{O}_*(Z_2)/J)$ contains no elements of order 4 either. From (4.5) we conclude

(4.6) <u>Corollary</u>: <u>For any</u> $p \geq 0$,

$$2\text{-}\underline{\text{rank}}\ (\mathrm{Tor}(\mathcal{O}_{4p}(Z_2)/J_{4p})) - 2\text{-}\underline{\text{rank}}\ (\mathrm{Tor}(\mathcal{O}_{4p}(Z_2))$$

$$= \underline{\text{rank}}\ (\Omega_{4p}) - 1.$$

This measures the excess of the torsion in $\mathcal{O}_{4p}(Z_2)/J_{4p}$ over the torsion in $\mathcal{O}_{4p}(Z_2)$. It is proved as follows. The excess 2-torsion arises from $J_{4p} \cap 2\mathcal{O}_{4p}(Z_2)$. More precisely from the $[M^{4p}]$ in Ω_{4p}/Tor such that

1) index $[M^{4p}] = 0 \bmod 2$
2) $[M^{4p}] \neq 0 \bmod 2$.

Since Ω/Tor is a polynomial ring over Z

$$\text{rank}\,\Omega_{4p} = \dim(\Omega_{4p}/\text{Tor}) \otimes Z_2.$$

We subtract off 1 because ind $[CP(2)]^p = 1$ for all $p \geq 0$.

(4.7) <u>Lemma</u>: <u>If</u> $a \in \mathcal{A}$ <u>then there is a</u> $b \in \mathcal{O}_*(Z_2)$ <u>for which</u>

$$2(a - j_*(b)) = 0.$$

Since (4.1) and (3.5) together prove that j_* induces an isomorphism

$$(\mathcal{O}_*(Z_2)/J)/\text{Tor} \simeq \mathcal{A}/\text{Tor}$$

the lemma is immediate.

(4.8) <u>Theorem</u>: <u>Any torsion class in</u> $\Omega_*(Z_2)$ <u>is the image under</u> ∂_* <u>of an element of order</u> 2 <u>in</u> $\mathcal{A}$.

We simply use (4.7) for $2(a-j_*(b)) = 0$ and $\partial_*(a-j_*(b)) = \partial_*\, a$. Remember Rosenzweig showed that im (∂_*) is precisely the 2-torsion in $\Omega_*(Z_2)$. With (4.6) and (4.8) it is possible in principal to compute the 2-rank of Tor $\mathcal{O}_*(Z_2)$. For any m we can use the isomorphism

$$\mathcal{A}_m \simeq \sum_{n+2k=m} \tilde{\Omega}_{n+1}(M(\gamma))$$

to determine Tor$(\mathcal{A}_m)$. Now $\Omega_{m-1}(Z_2) \simeq \Omega_{m-1} \oplus \tilde{\Omega}_{m-1}(Z_2)$ and it is known that $\tilde{\Omega}_{m-1}(Z_2) \simeq \mathcal{N}_{m-2}$. If $m \neq 0 \bmod 4$ then 2-rank $\mathcal{O}_m(Z_2)$ = 2-rank $\mathcal{A}_m$ - 2-rank Ω_{m-1} - dim $\mathcal{N}_{m-2}$.

While if m = 4p

$$2\text{-rank } \mathcal{O}_{4p}(Z_2) = 2\text{-rank } \mathcal{A}_{4p} - 2\text{-rank } \Omega_{4p-1} - \dim \mathcal{N}_{4p-2} - \text{rank } \Omega_{4p} + 1.$$

While the actual numbers are not meaningful by themsevles we did want to point out that the 2-rank of $\mathcal{O}_*(Z_2)$ can be determined in every dimension.

5. A trace invariant

Professors Atiyah and Hirzebruch pointed out to us the form that the Atiyah-Bott Fixed Point Theorem takes for orientation preserving involutions on closed oriented manifolds. In fact we shall derive the appropriate formula as a corollary of (4.1), but the Trace invariant has additional application to which we shall draw the reader's attention.

Consider a triple $(T, V, (\cdot,\cdot))$ wherein T is a real linear transformation of period 2 on a finite dimensional real vector space V equipped with a bilinear, symmetric innerproduct (v,w) which is non-singular in the sense that if $v \neq 0$ then there is a w with $(v,w) \neq 0$. We also suppose that $(Tv,Tw) \equiv (v,w)$ of course.

Denote by $\mathcal{L}(T,V)$ the algebra of all linear transformations of V into itself which commute with T and denote by $GL(T,V) \subset \mathcal{L}(T,V)$ those which are also invertible. Fix a positive definite symmetric bilinear inner-product $\langle v,w \rangle$ for which $\langle Tv,Tw \rangle \equiv \langle v,w \rangle$. Denote by $\mathcal{S}y(T) \subset \mathcal{L}(T,V)$ the vector space of those linear operators commuting with T which are symmetric; that is, $\langle v,Lw \rangle \equiv \langle Lv,w \rangle$, and by $\mathcal{S}y_+(T) \subset \mathcal{S}y(T)$ those which in addition are positive definite; that is, $\langle v,Lv \rangle > 0$ if $v \neq 0$. Similarly $\mathcal{D}(T) \subset \mathcal{L}(T,V)$ is the vector space of those linear operators which commuted with T and satisfy

$$(v,Dw) \equiv (Dv,w)$$

while $\mathcal{D}_+(T) \subset \mathcal{D}(T)$ is the subset of D for which in addition $(v,Dv) > 0$ if $v \neq 0$. Observe that $T \in \mathcal{D}(T) \cap \mathcal{S}y(T)$ for

$$(v,Tw) = (TTv,Tw) = (Tv,w)$$

and similarly for $\langle \cdot,\cdot \rangle$.

We shall now exhibit a linear isomorphism of $\mathcal{S}y(T)$ with $\mathcal{D}(T)$ which makes $\mathcal{S}y_+(T)$ onto $\mathcal{D}_+(T)$. Using the non-singularity of $(\cdot,\cdot)$ we argue by duality that for each $w \in V$ there is a unique $D_o w \in V$ such that $(v,D_o w) = \langle v,w \rangle$ for all $v \in V$. It follows easily that D_o is linear. Now

$$(v,D_o w) = \langle v,w \rangle = \langle w,v \rangle = (w,D_o v) = (D_o v,w), \; (v,D_o v) = \langle v,v \rangle > 0$$

if $v \neq 0$.

It will follow that $D_o \in \mathcal{D}_+(T)$ if we can show $D_o T = TD_o$, but

$$(v,D_o Tw) = \langle v,Tw \rangle = \langle Tv,w \rangle = (Tv,D_o w) = (v,TD_o w)$$

for all v and w so the commutativity follows from the non-singularity of $(\cdot,\cdot)$. If $L \in \mathcal{S}y(T)$, then $D_o L \in \mathcal{D}(T)$ for

$$(v,D_o Lw) = \langle v,Lw \rangle = \langle w,Lv \rangle = (w,D_o Lv) = (D_o Lv,w),$$

and $(v,D_o Lv) = \langle v,Lv \rangle$, hence $D_o L \in \mathcal{D}_+(T)$ if and only if $L \in \mathcal{S}y_+(T)$. Next, if $D \in \mathcal{D}(T)$ then for any $w \in V$ there is a unique Lw for which $(v,Dw) = \langle v,Lw \rangle$ for all v. Again L is linear, symmetric and commutes with T, so $L \in \mathcal{S}y(T)$. Note that

$$(v,D_o Lw) = \langle v,Lw \rangle = (v,Dw),$$

hence $D_oL = D$ and so we have

$$D_o: \mathcal{L}_y(T) \simeq \mathcal{D}(T).$$

It is trivial to see that $\mathcal{L}_{y_+}(T)$, and hence $\mathcal{D}_+(T)$, is convex. We can argue that $\mathcal{L}_{y_+}(T)$ is open in $\mathcal{L}_y(T)$ in the following standard manner. Suppose $L \in \mathcal{L}_{y_+}(T)$ then, using the compactness of the unit sphere, there is $\epsilon > 0$ such that $\langle v,Lv \rangle \geq \epsilon$ for all $\langle v,v \rangle = 1$. If $L' \in \mathcal{L}_y(T)$ and $|\langle v,(L-L')v \rangle| < \epsilon$ for all $\langle v,v \rangle = 1$ it follows $L' \in \mathcal{L}_{y_+}(T)$ also.

We have shown that $\mathcal{D}_+(T)$ is an open, non-empty, convex subset of the linear space $\mathcal{D}(T)$. Thus $\mathcal{D}_+(T)$ is homeomorphic to $\mathcal{D}(T)$ itself. We are going to define a map of period 2 on $\mathcal{D}_+(T)$ and apply the Smith fixed point theorem to show that the fixed point set is non-empty and connected.

We claim that if $D \in \mathcal{D}_+(T)$ then $D^{-1} \in \mathcal{D}_+(T)$ also because

$$(v,D^{-1}w) = (DD^{-1}v,D^{-1}w) = (D^{-1}v,w)$$

and

$$(v,D^{-1}v) = (DD^{-1}v,D^{-1}v) = (D^{-1}v,D(D^{-1}v)) > 0 \text{ if } v \neq 0.$$

The transformation $D \longrightarrow D^{-1}$ is a map of period 2 on the open convex set $\mathcal{D}_+(T)$ and it has a connected non-empty fixed point set. Obviously $D = D^{-1}$ if and only if $D^2 = I$.

(5.1) Lemma: The set of all linear operators on V

<u>satisfying</u>

1) $DT = TD$

2) $D^2 = I$

3) $(v,Dw) = (Dv,w)$

4) $(v,Dv) > 0$ if $v \neq 0$

form a non-empty connected subset of $GL(T,V)$ and any two such operators are conjugate in $GL(T,V)$.

The reader will see that in what follows it is immaterial which D in (5.1) is used. Choose D and let $V = V_+ \oplus V_-$ be the ±1-eigenspace decomposition of V under D. The inner-product is positive definite on V_+ and negative definite on V_-. Since $DT = TD$ the eigenspaces V_+ and V_- are T-invariant. We put

$$Tr(T,V,(\cdot,\cdot)) = \text{trace } (T,V_+) - \text{trace } (T,V_-).$$

Recall that

$$\text{index } (V,(\cdot,\cdot)) = \dim V_+ - \dim V_-.$$

Now use T to decompose V_+, V_- as

$$V_+ = V_+^e \oplus V_+^o$$

$$V_- = V_-^e \oplus V_-^o.$$

The e(ven) denotes the +1 eigenspace of T and the o(dd) the -1. Then

$$Tr(T,V,(\cdot,\cdot)) = (\dim V_+^e - \dim V_+^o) - (\dim V_-^e - \dim V_-^o)$$

while

$$\text{ind}\ (V,(\bullet,\bullet)) = \dim V_+^e + \dim V_+^o - (\dim V_-^e + \dim V_-^o)$$

thus

$$\text{Tr}(T,V,(\bullet,\bullet)) - \text{ind}\ (V,(\bullet,\bullet)) = 2(\dim V_-^o - \dim V_+^o) = -2\text{ind}\ (V^o,(\bullet,\bullet)).$$

(5.2) Lemma: *For any* $(T,V,(\bullet,\bullet))$

$$\text{Tr}(T,V,(\bullet,\bullet)) \equiv \text{ind}\ (V,(\bullet,\bullet))$$

modulo 2.

Given $(T,V,(\bullet,\bullet))$ and $(T',V',(\bullet,\bullet)')$ let $(T,V,(\bullet,\bullet)) \oplus (T',V',(\bullet,\bullet)')$ be $V + V'$ with innerproduct

$$((v+v',w+w')) = (v,w) + (v',w')'$$

and $(T+T')(v+v') = Tv + T'v'$. Similarly, $(T,V,(\bullet,\bullet)) \otimes (T',V'(\bullet,\bullet)')$ gives to $V \otimes V'$ the innerproduct

$$((v \otimes v',w \otimes w')) = (v,w)(v',w')'$$

and $T \otimes T'$. Using the definitions the reader may show

(5.3) Lemma: *For any pair* $(T,V,(\bullet,\bullet))$ *and* $(T',V',(\bullet,\bullet)')$

$$\text{Tr}((T,V,(\bullet,\bullet)) \oplus (T',V',(\bullet,\bullet)') = (\text{Tr}(T,V,(\bullet,\bullet))) + (\text{Tr}(T',V',(\bullet,\bullet)'))$$

and

$$\text{Tr}((T,V,(\bullet,\bullet)) \otimes (T',V',(\bullet,\bullet)')) = \text{Tr}(T,V,(\bullet,\bullet))\ \text{Tr}(T',V',(\bullet,\bullet)').$$

We also observe that $\text{Tr}(T,V,(\bullet,\bullet)) = 0$ if there is a T-invariant subspace $W \subset V$ with $2 \dim W = \dim V$ and on which the inner-product is totally degenerate. The assumption of degeneracy implies $W \cap V_+ = W \cap V_- = \{0\}$ and by the dimensional consideration we see the projections induce isomorphisms $(T,V_+) \cong (T,W) \cong (T,V_-)$.

We can use this to define a ring homomorphism $Tr : \mathcal{O}_*(Z_2) \to Z$. We agree Tr is trivial on $\mathcal{O}_m(Z_2)$ if $m \neq 0$ mod 4. For (T,M^{4p}) we define the usual inner-product on $H^{2p}(M^{4p};R) = V$ by $(v,w) = \langle v \cup w, \sigma_{4p} \rangle \in R$, where $\sigma_{4p} \in H_{4p}(M^{4p};Z)$ is the orientation class. Since T preserves orientation $(T^*v,T^*w) = (v,w)$, thus we have $(T^*,V,(\cdot,\cdot))$. We set

$$Tr(T,M^{4p}) = Tr(T^*,H^{2p}(M^{4p};R),\ (\cdot,\cdot)).$$

The usual argument will show Tr only depends on the bordism class $[T,M^{4p}]$. First, if σ_{4p} is replaced by $-\sigma_{4p}$ the result is $-Tr(T,M^{4p})$. If there is a (T,B^{4p+1}) with $(T,\partial B^{4p+1}) = (T,M^{4p})$ then the image of

$$i^*: (T^*,H^{2p}(B^{4p+1};R)) \to (T^*,H^{2p}(M^{4p};R))$$

is the T^*-invariant subspace $W \subset H^{2p}(M^{4p};R)$ on which the inner-product is totally degenerate and for which $2 \dim W = \dim H^{2p}(M^{4p};R)$. With (5.3) and the usual arguments for the case of index it can be shown that $Tr : \mathcal{O}_*(Z_2) \to Z$ is a ring homomorphism. Finally note that if $T = I$ is the trivial involution then $Tr[T,M^{4p}] = \text{index}\,[M^{4p}]$.

(5.4) Lemma: If $[M^{4p}] \in \Omega_{4p}$ then $Tr([Z_2,Z_2][M^{4p}]) = 0$.

Note that

$$H^{2p}(Z_2 \times M^{4p};R) \cong H^{2p}(M^{4p};R) \oplus H^{2p}(M^{4p};R)$$

and that $T^*(v+v') = v' + v$. The inner-product on $H^{2p}(Z_2 \times M^{4p};R)$ is $((v+v',w+w')) = (v,w) + (v',w')$. Thus if $H^{2p}(M^{4p};R) = V_+ \oplus V_-$ then we have

$$H^{2p}(Z_2 \times M^{4p};R) = (V_+ \oplus V_+) \oplus (V_- \oplus V_-)$$

and trace $(T^*, V_+ \oplus V_+) = 0 =$ trace $(T^*, V_- \oplus V_-)$.

(5.5) <u>Corollary:</u> <u>If</u> $[Z_2,Z_2][M^{4p}]$ <u>lies in</u> $2\mathcal{O}_{4p}(Z_2)$ <u>then</u> <u>index</u> $[M^{4p}] = 0$

If $2[T,X^{4p}] = [Z_2,Z_2][M^{4p}]$ then $2([X^{4p}] - [M^{4p}]) = 0$ so ind $[X^{4p}] =$ ind $[M^{4p}]$. By (5.4) 2 Tr $[T,X^{4p}] = 0$, so Tr $[T,X^{4p}] = 0$, but as we noted in (5.2), this implies

$$\text{ind}[X^{4p}] = \text{ind}[M^{4p}] \equiv 0$$

modulo 2. This completes the proof of theorem (4.5) of course.

In any case we see that Tr induces a ring homomorphism

$$\text{Tr} : \mathcal{O}_*(Z_2)/J \longrightarrow Z.$$

We also know that $\mathcal{S} : \mathcal{O}_*(Z_2) \longrightarrow \mathfrak{M}$ induces an isomorphism $\mathcal{O}_*(Z_2)/J \otimes Z(1/2) \simeq \mathfrak{M} \otimes Z(1/2)$. Thus we ought to be able to find a formula for Tr in terms of the self-intersection of the fixed point set with itself. Define

$$\ell : \Omega_{4q}(BO(2p)) \longrightarrow Z$$

by assigning to $[\xi \longrightarrow V^{4q}]$ the integer $(-1)^p$ index $[V^{4q}]$.

This extends uniquely to a ring homomorphism

$$\ell : \mathcal{M} \to Z.$$

(5.6) <u>Theorem:</u> <u>The homomorphism</u> Tr : $\mathcal{O}_*(Z_2) \to Z$ <u>is equal to the composite homomorphism</u>

$$\mathcal{O}_*(Z_2) \xrightarrow{\mathcal{S}} \mathcal{M} \xrightarrow{\ell} Z.$$

Since by (4.1), $(\mathcal{O}_*(Z_2)/J) \otimes Z(1/2)$ is the polynomial ring over $\Omega \otimes Z(1/2)$ generated by $\{[A, CP(2p)]\}_{p=1}^{\infty}$ where

$$A[z_1, \ldots, z_{2p+1}] = [-z_1, z_2, \ldots, z_{2p+1}]$$

it is enough to verify the result for these examples. Since $(A^*, H^{2p}(CP(2p), R))$ is the identity $\text{Tr}[A, CP(2p)] = 1$. As we have seen, $\mathcal{S}[A, CP(2p)] = -[\eta_{2p-2} \to CP(2p-2)] \in \Omega_{4p-4}(BO(2))$ and $\ell(-[\eta_{2p-2} \to CP(2p-2)]) = \text{ind}[CP(2p-2)] = 1$.

(5.7) <u>Corollary:</u> <u>If</u> (T, M^{4p}) <u>is an orientation preserving involution on a closed oriented manifold for which every component of the fixed point set has dimension less than</u> 2p+1, <u>then</u>

$$2[T, M^{4p}] = [Z_2, Z_2][M^{4p}]$$

<u>if and only if</u> $\text{Tr}[T, M^{4p}] = 0$.

From elementary dimensional considerations $\mathcal{S}[T, M^{4p}] = s[\xi \to F^{2p}; \phi] \in \Omega_0(BO(2p)) \simeq Z$. The last isomorphism is given by $(-1)^p \ell$. We apply (4.2).

To complete the section we discuss the relation of $\mathcal{O}_*(Z_2)$

to $I_*(Z_2)$, the unoriented algebra of bordism classes of involutions on closed manifolds, [6 Ch.IV]. Every element of $I_*(Z_2)$ has order 2 and there is the forgetful homomorphism which neglects orientation. The analog for $\mathcal{A}$ in the unoriented case is

$$\mathcal{M}_m = \sum_{n+k=m} \mathcal{N}_n(BO(k)).$$

With $\mathcal{M} = \sum_0^\infty \mathcal{M}_m$ there is a short exact sequence, [6] ,

$$0 \longrightarrow I_*(Z_2) \xrightarrow{j_*} \mathcal{M} \xrightarrow{\partial_*} \mathcal{N}_*(Z_2) \rightarrow 0.$$

From general bordism theory, [6, Ch. II], the sequence

$$\tilde{\Omega}_{n+1}(M(\gamma)) \xrightarrow{2} \tilde{\Omega}_{n+1}(M(\gamma)) \xrightarrow{r} \tilde{\mathcal{N}}_{n+1}(M(\gamma))$$

is exact, and there is the Thom isomorphism

$$\mathcal{N}_n(BO(2k)) \simeq \tilde{\mathcal{N}}_{n+1}(M(\gamma))$$

therefore the sequence

$$\mathcal{A} \xrightarrow{2} \mathcal{A} \xrightarrow{r} \mathcal{M}$$

is exact, where r neglects the orientation of $\zeta \oplus \tau$.

(5.8) Theorem: If $m \neq 0 \bmod 4$ then

$$r : \mathcal{O}_m(Z_2) \rightarrow I_m(Z_2)$$

is a monomorphism. If $b \in \mathcal{O}_{4p}(Z_2)$ lies in the kernel of r, then $b = 2b' + [Z_2, Z_2][M^{4p}]$ for some $b' \in \mathcal{O}_{4p}(Z_2)$ and $[M^{4p}] \in \Omega_{4p}/\text{Tor}$.

<u>Finally</u>, $b \in 2\mathcal{O}_{4p}(Z_2)$ <u>if and only if</u>

1) $r(b) = 0$

2) $Tr(b) \equiv$ <u>index</u> (b) <u>mod</u> 4.

In general we have

$$\begin{array}{ccccc} \mathcal{O}_m(Z_2) & \xrightarrow{2} & \mathcal{O}_m(Z_2) & \xrightarrow{r} & I_m(Z_2) \\ \downarrow j_* & & \downarrow j_* & & \downarrow j_* \\ \mathcal{A} & \xrightarrow{2} & \mathcal{A} & \xrightarrow{r} & \mathcal{M} \end{array}$$

If $r(b) = 0$, $b \in \mathcal{O}_m(Z_2)$ then $rj_*(b) = 0 \in \mathcal{M}_m$ and there is $a \in \mathcal{A}_m$ with $2a = j_*(b)$. By (4.7) there is $b' \in \mathcal{O}_m(Z_2)$ with

$$2(a - j_*(b')) = 0 = j_*(b) - 2j_*(b').$$

Since $\mathcal{O}_m(Z_2) \to \mathcal{A}_m$ is a monomorphism for $m \neq 0 \bmod 4$ we have $2b' = b$. But $\mathcal{O}_m(Z_2)$ consists entirely of elements of order 2, hence $b = 0$ and the first part of (5.8) is established.

If $m = 4p$ then since $j_*(b - 2b') = 0$ we have $b = 2b' + [Z_2, Z_2][M^{4p}]$. Now $b \in 2\mathcal{O}_*(Z_2)$ if and only if $[Z_2, Z_2][M^{4p}] \in 2\mathcal{O}_*(Z_2)$; that is, by (4.5) if and only if

$$\text{ind}\,[M^{4p}] = 0 \bmod 2.$$

But

$$\text{ind}\,(b) = 2\,\text{ind}\,(b') + 2\,\text{ind}\,[M^{4p}]$$

and by (5.4)

$$Tr(b) = 2Tr(b')$$

so that

$$\text{ind }(b) - \text{Tr}(b) = 2 \text{ ind } [M^{4p}] + 2(\text{ind }(b') - \text{Tr}(b'))$$

or

$$\text{ind } [M^{4p}] = 1/2\ (\text{ind }(b) - \text{Tr}(b)) - (\text{ind }(b') - \text{Tr}(b')).$$

Since index $\equiv$ Tr mod 2 we see that index $[M^{4p}] \equiv 0$ mod 2 if and only if ind (b) $\equiv$ Tr(b) mod 4.

This last be summarized into a Rochlin sequence as follows. Define

$$\lambda : \mathcal{O}_{4p}(Z_2) \longrightarrow Z_2$$

by $\lambda[T, M^{4p}] = 1/2\ (\text{ind } [M^{4p}] - \text{Tr } [T, M^{4p}])$ then

(5.9) <u>Corollary</u>: <u>For</u> $p \geq 0$ <u>the sequence</u>

$$\mathcal{O}_{4p}(Z_2) \xrightarrow{2} \mathcal{O}_{4p}(Z_2) \xrightarrow{r+\lambda} I_*(Z_2) \oplus Z_2$$

<u>is exact</u>.

(5.10) <u>Corollary</u>: <u>If</u> $2b = 0$ <u>and</u> $r(b) = 0$, <u>then</u> $b = 0$.

We need only consider $b \in \mathcal{O}_{4p}(Z_2)$. Since $2b = 0$, $\text{Tr}(b) = \text{ind }(b) = 0$, so since $r(b) = 0$ also, b is divisible by 2. There are no elements of order 4 in $\mathcal{O}_*(Z_2)$.

We can describe the structure of $\mathcal{O}_*(Z_2)/\text{Tor}$ as an algebra over Ω/Tor. Consider a graded polynomial ring $\Omega/\text{Tor}\,[K_0, K_4, \ldots, K_{4p}, \ldots]$. Let $\mathcal{J}$ be the ideal generated by $(K_0(K_0-2),\ K_0(K_4-[CP(2)]), \ldots K_0(K_{4p}-[CP(2p)]), \ldots)$.

(5.12) <u>Theorem</u>: <u>There is a natural isomorphism of</u> $\mathcal{O}_*(Z_2)/\text{Tor}$ <u>with the quotient of the polynomial ring</u>

Ω/Tor $[K_o,\ldots,K_{4p},\ldots]$ <u>by</u> <u>the</u> <u>ideal</u> $\mathcal{J}$.

Send $K_o \longrightarrow [Z_2, Z_2]$ and $K_{4p} \longrightarrow [A, CP(2p)]$ for $p > 0$.

6. Examples

The first example which comes to mind is $(T, V^{2p} \times V^{2p})$ where V^{2p} is a closed oriented manifold, the product manifold receives the product involution and $T(x,y) = (y,x)$. We assert that $\operatorname{Tr}[T, V^{2p} \times V^{2p}] = \chi(V^{2p})$. The fixed point set is the diagonal, and the normal bundle is the tangent bundle to V^{2p}. In section 2, p. 90, we saw that if $\tau \oplus \tau \to V^{2p}$ is given the orientation arising from the complex structure then $S[\tau \to V^{2p}, \theta_c] = \chi(V^{2p})$. However, in this case we are to use the product orientation on $\tau \oplus \tau \to V^{2p}$ which imparts to the self-intersection the sign adjustment $(-1)^p \chi(V^{2p})$, but application of ℓ multiplies by another $(-1)^p$, hence $\operatorname{Tr}[T, V^{2p} \times V^{2p}] = \chi(V^{2p})$ as asserted. Note explicitly that $[T, V^{2p} \times V^{2p}]$ does not depend on $[V^{2p}]$ alone! In fact, by (5.7), $[T, V^{2p} \times V^{2p}] = 0$ if and only if $\chi(V^{2p}) = 0$ and $2[V^{2p}] = 0 \in \Omega_{2p}$.

Another example arises from a conjugation involution on an almost complex manifold M^{4p}, [6, Ch. IV]. The tangent bundle has an orthogonal bundle map

$$\begin{array}{ccc} \tau & \xrightarrow{J} & \tau \\ & \searrow \quad \swarrow & \\ & M^{4p} & \end{array}$$

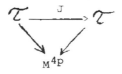

with $J^2 = -I$. This complex structure orients $\tau \to M^{4p}$ canonically. We suppose that we are given an involution (T, M^{4p}) for which JdT = -dTJ. We call this a conjugation. If we regard $(\tau, J) \to M^{4p}$ as a complex bundle over M^{4p}, then the bundle induced by $T : M^{4p} \to M^{4p}$ from (τ, J) is $(\tau, -J) \to M^{4p}$; that is, the conjugate bundle. It is important to realize that since the complex dimension is even, J and -J induce the same orientation so that T is orientation preserving in this case. This is merely the observation that on C^{2p} the operation of conjugation is orientation preserving.

Come down to dT_x, J_x and τ_x at a fixed point $x \in M^{4p}$. The eigenspaces of dT_x split τ_x into $\tau'_x \oplus \zeta_x$, which are respectively the ±1 eigenspaces. The fixed vectors in τ'_x are the tangent vectors to the fixed point set at x, while the skew vectors in ζ_x are normal to the fixed point set. Since $dT_x J_x = -J_x dT_x$ we see $J : \tau' \to \zeta$ is an orthogonal bundle equivalence of τ' with ζ, thus the fixed point set is a 2p-manifold $F^{2p} \subset M^{4p}$ and so is called the real fold of the conjugation. The restriction of τ to F^{2p} has the natural form $J\tau' \oplus \tau' = \zeta \oplus \tau'$. This says that the orientation on $\tau' \oplus \tau' = \zeta \oplus \tau'$ given by $(v,w) \to (-w,v)$ co-incides with the orientation of $\tau \to F^{2p}$ given by -J, and this is the natural orientation inherited from that of M^{4p}. Now

$$s[\tau' \to F^{2p}, \sigma_c] = \chi(F^{2p})$$

but $\rho s[\tau' \to F^{2p}, \sigma_c] = (-1)^p \chi(F^{2p})$.

(6.1) Lemma: *If (T, M^{4p}) is a conjugation on a closed* almost complex manifold, then

$$\mathrm{Tr}\,[T,M^{4p}] = (-1)^p\chi(F^{2p}).$$

To verify this consider the conjugation $(T,CP(2p))$ given by $T[z_1,\ldots,z_{2p+1}] = [\bar{z}_1,\ldots,\bar{z}_{2p+1}]$. Now $T^*(c) = -c$ for the generator $c \in H^2(CP(2p);Z)$, hence $T^*(c^p) = (-1)^p c^p$. Since the innerproduct on $H^{2p}(CP(2p);R)$ is positive definite, $\mathrm{Tr}[T,CP(2p)] = (-1)^p$. The real fold is $RP(2p)$, so $(-1)^p\chi(RP(2p)) = (-1)^p$.

Continuing with a conjugation, denote by $\tilde{F}^{2p} \to F^{2p}$ the canonically oriented orientation double covering of the real fold and by $(\tilde{T},\tilde{F}^{2p}\times\tilde{F}^{2p})$ the orientation preserving involution $(x,y) - (y,x)$

(6.2) Theorem: If (T,M^{4p}) is a conjugation on a closed almost complex manifold, then

$$2\,[T,M^{4p}] + (-1)^{p+1}[\tilde{T},\tilde{F}^{2p}\times\tilde{F}^{2p}] = [Z_2,Z_2][M^{4p}].$$

Recall $\mathrm{Tr}[T,M^{4p}] = (-1)^p\chi(F^{2p})$ and $\mathrm{Tr}[\tilde{T},\tilde{F}^{2p}\times\tilde{F}^{2p}] = \chi(\tilde{F}^{2p}) = 2\chi(F^{2p})$. We can apply (5.7) to $2[T,M^{4p}] + (-1)^{p+1}[\tilde{T},\tilde{F}^{2p}\times\tilde{F}^{2p}]$, but $[\tilde{F}^{2p}] = 0$ in Ω_{2p} since $\tilde{F}^{2p}$ admits an orientation reversing involution without fixed points, hence with (5.7)

$$2(2[T,M^{4p}] + (-1)^{p+1}[\tilde{T},\tilde{F}^{2p}\times\tilde{F}^{2p}] - [Z_2,Z_2][M^{4p}]) = 0.$$

If we can show this torsion class is 0 in $I_{4p}(Z_2)$ then by (5.10) we are done.

On the face of it, it is enough to show $[\tilde{T},\tilde{F}^{2p}\times\tilde{F}^{2p}]_2 = 0$. Let $d : \tilde{F}^{2p} \to \tilde{F}^{2p}$ be the orientation reversing fixed point free double covering involution. Then $d \times d$ on $\tilde{F}^{2p}\times\tilde{F}^{2p}$ is a fixed point free (now orientation preserving) involution which commutes with $\tilde{T}$. Extend $\tilde{T}$ to the mapping cylinder of the quotient $\tilde{F}^{2p}\times\tilde{F}^{2p} \to (\tilde{F}^{2p}\times\tilde{F}^{2p})/(d\times d)$ to see that $[T,F^{2p}\ \ F^{2p}]_2 = 0$ in $I_{4p}(Z_2)$.